M000220292

Practical Analytical Chemistry, Lab Manual

Sintayehu L. Kitaw

Practical Analytical Chemistry, Lab Manual

LAP LAMBERT Academic Publishing

Impressum / Imprint

Bibliografische Information der Deutschen Nationalbibliothek: Die Deutsche Nationalbibliothek verzeichnet diese Publikation in der Deutschen Nationalbibliografie; detaillierte bibliografische Daten sind im Internet über http://dnb.d-nb.de abrufbar.

Alle in diesem Buch genannten Marken und Produktnamen unterliegen warenzeichen-, marken- oder patentrechtlichem Schutz bzw. sind Warenzeichen oder eingetragene Warenzeichen der jeweiligen Inhaber. Die Wiedergabe von Marken, Produktnamen, Gebrauchsnamen, Handelsnamen, Warenbezeichnungen u.s.w. in diesem Werk berechtigt auch ohne besondere Kennzeichnung nicht zu der Annahme, dass solche Namen im Sinne der Warenzeichen- und Markenschutzgesetzgebung als frei zu betrachten wären und daher von jedermann benutzt werden dürften.

Bibliographic information published by the Deutsche Nationalbibliothek: The Deutsche Nationalbibliothek lists this publication in the Deutsche Nationalbibliografie; detailed bibliographic data are available in the Internet at http://dnb.d-nb.de.

Any brand names and product names mentioned in this book are subject to trademark, brand or patent protection and are trademarks or registered trademarks of their respective holders. The use of brand names, product names, common names, trade names, product descriptions etc. even without a particular marking in this work is in no way to be construed to mean that such names may be regarded as unrestricted in respect of trademark and brand protection legislation and could thus be used by anyone.

Coverbild / Cover image: www.ingimage.com

Verlag / Publisher:
LAP LAMBERT Academic Publishing
ist ein Imprint der / is a trademark of
OmniScriptum GmbH & Co. KG
Heinrich-Böcking-Str. 6-8, 66121 Saarbrücken, Deutschland / Germany
Email: info@lap-publishing.com

Herstellung: siehe letzte Seite /
Printed at: see last page
ISBN: 978-3-659-29045-9

Table of Contents

Preface

This Analytical Chemistry Laboratory Manual is written for use to undergraduate chemistry students and their instructors. It is also helpful for non-chemistry majors taking analytical chemistry laboratory courses. With this in mind, the experiments to be performed are dealt in detail with brief introduction, procedures, data analysis instruments and discussions. The author was intended to cover the topics discussed in undergraduate analytical chemistry courses. It is tried to organize the manual in two modules. Module I focuses on volumetric and gravimetric methods of chemical analysis while module II deals with instrumental methods of chemical analysis.

In the classical or wet procedure (volumetric and gravimetric method), no sophisticated mechanical or electronic instruments are required. Simple glassware such as beakers, test tubes, pipettes, burettes, measuring cylinders etc and a balance are sufficient. In this procedure, stoichiometries of chemical reactions are used to perform the analysis. As an example, the levels of calcium in a solution can be quantitatively analyzed by adding excess oxalate ions at high pH. The resulting calcium oxalate precipitate is separated by filtration followed by igniting and weighing. From the weighed mass of calcium oxalate, the amount of calcium in the sample could be determined from the reaction stoichiometry:

$$Ca^{2+} + C_2O_4^{2-} + H_2O \rightarrow CaC_2O_4.H_2O(s)$$

An analysis of this kind which depends on measurement of mass of a solid product is called gravimetric analysis.

A different aspect of classical method of analysis is volumetric analysis which involves titration of a solution containing an analyte with a standard solution of accurately measured volume. This is achieved by locating the equivalent point of the titration using appropriate indicators. In the determination of acetic acid in vinegar solution, for example, a measured volume of vinegar solution is titrated against a standard solution of sodiumhydroxide, NaOH, until the equivalent point is signaled by the change of color of indicator. Since stoichiometrically equivalent amounts of the acid and base react at the equivalent point, the amount of acetic acid in vinegar is calculated using the known amount of NaOH from the reaction stoichiometry:

$$CH_3COOH + NaOH \rightarrow CH_3COONa + H_2O$$

Now a days, in most analytical chemistry laboratories, chemical analyses are usually performed by using instrumental methods. Instrumental methods of chemical analysis rely upon instruments of some kind either to make a critical measurement during an analysis or to perform the entire analysis. Instrumental methods of analysis are used to determine very small quantities of an analyte with better accuracy and precision than classical methods. Some divisions of instrumental analysis include optical methods, chromatographic methods, electroanalytical methods, thermal methods, kinetic methods and magnetic resonance methods. In some cases of instrumental analysis, the analyses are more considerably automated where results are displayed by a group of inter-connected instruments.

Writing a manual is a difficult and lengthy process which needs the co-operation of many people. The author would acknowledge the assistance and extremely helpful comments of Mr Yihenew Simegniew, Head of department of chemistry, in Debre Markos University. The patience of Mrs Anna Voronina in timely and frequently communicating with the author and helping sending book project editing guide and pdf converter soft-ware made the publication of the manual possible.

Sintayehu L. Kitaw

LABORATORY REPORT WRITING

Chemistry is the study of matter, its properties, and changes associated with matter. In chemistry classrooms, students will learn a great deal of the information that has been gathered by scientists about matter. But, chemistry is just beyond information. It is also a process for finding out more about matter and its changes. Laboratory activities are the primary means that chemists use to learn more about matter. The activities in this Laboratory Manual require learners to form and test hypotheses, measure and record data and observations, analyze those data, and draw conclusions based on those data and their knowledge of chemistry. These processes are the same as those used by professional chemists and all other scientists.

Understanding of the techniques, concepts, and calculations covered in the laboratory course will provide the foundation for future chemistry and science courses and future work and thinking after the university experience. Basic techniques such as mass measurement, volume transfer, solution preparation and dilution, titration, and qualitative analysis must be done safely balancing precision and speed. Concepts must then be applied to collected data to calculate final results using unit cancellation, often called dimensional analysis, and determine the quality or precision of the data expressed using proper significant figures, graphical analysis, and statistics. To excel in the analytical chemistry laboratory courses, students must demonstrate mastery of all the key techniques and related concepts and calculations.

The data and results for each experiment are placed in a data table and will be evaluated to determine if techniques, concepts, and calculations have been understood. If they use the equipment properly, record observations accurately, and calculate correctly, they will obtain an acceptable answer (close to the true value). Each experiment will have pre and post lab work that will be turned in as part of the report sheet or will be done via a web based system. The pre lab preparation for each experiment will be tested by a pre lab quiz and overall course mastery by a lab practical.

While performing experiments, we make observations, collect and analyze data, and formulate generalizations about the data. When students work in the laboratory, they should record all data in a laboratory report. An analysis of data is easier if all data are recorded in an organized, logical manner. Tables and graphs are often used for this purpose. When writing a laboratory report:

- Always write in third person

- Write in full sentences except for the materials list, and check spelling and grammar.
- Use significant figures and units regarding measurements and calculations
- Neatness is important. Use rulers when needed (especially when using tables and graphs), type if possible
- Do not copy verbatim (word for word) from the lab handout or any other source.

The report should have the following sections.

1. **Title**- Heading, name of group members, section, teacher name, lab assistant name, date of lab report submission. All these information should appear on front cover of the lab report.

2. **Introductory Paragraph** – This section should be written in complete sentences and should connect lab concepts to theoretical content. The introduction should provide background information on the history of the concepts tested, scientists, theories, and any laws tested in the experiment. Cite Sources Used. The introduction should contain any prior knowledge on which the experiment is based including an explanation of principles, definitions, experimental techniques, theories and laws.

3. **Objective/ Purpose:** The objective is a concise statement in complete sentences outlining the purpose of the experiment. The purpose section of a lab is where you tell the reader your reason for doing the lab in the first place and briefly summarize any relevant background information about the experiment, including any relevant chemical equations and/or algebraic equations.

4. **Materials:** (Bulleted List) the materials section is a list of all equipment, reagents (chemicals), and computer programs that were used to complete the experiment. Drawings of the apparatus setup should be included in this section if needed. The materials list must be complete. Indicate how much of each material will be used in the experiment. If you plan on arranging some of the equipment into a more complex setup (for example, if you are going to heat something over a Bunsen burner, you will need a ring stand, wire gauze, etc.), draw it as well as mention the equipment used.

5. **Procedure:** This section may be written in either paragraphs or numbered steps. Explain the test design, and allow for pictures and diagrams. The procedure is a detailed statement (step by step) of how the experiment was performed such that the experiment could be repeated using your report. Safety

precautions that were followed should be stated in this section. The procedure must be written in the impersonal (3rd person) past tense: e.g. do not say 'we are recording the absorbance at intervals of 10 nm wave lengths', instead say 'The absorbance was recorded at intervals of 10 nm wave lengths'.

6. **Data, Results and Observations**: This is a collection of observations, measurements, multiple trials, data tables, charts, and repeating steps. This section may consist of quantitative and/or qualitative observations of the experiment. A qualitative piece of data is a written description and/or sketch of what was seen during the experiment. Quantitative information may be in the form of a table or simply a written description. When graphs are required, special attention should be paid to the following items: the type of graph expected (straight line or curve), utilizing the entire graph paper, plotted point size, title of the graph, and axis labels. When numerous measurements have made, data is to be placed in a data table whenever possible. Figure headings are placed below the figure and should give a short description of the figure. The figure number should be in bold font. Table headings are found above the table and should also have a brief description.

7. **Data analysis and Calculations:** These include: graphs, error calculations, equations and statistical analysis. One example of each type of calculation should be included. Results from numerous calculations should be placed in a data table with the proper number of significant figures and correct units. % yield and % error calculations should be included when possible.

8. **Conclusion:** The conclusion is a concise statement that answers the objective. The result of percent error and/or percent yield should be discussed and compared with known results. A portion of the conclusion should be dedicated to error analysis which discusses any possible sources of error that may have contributed to the percent error or yield. List at least two possible errors in the lab, as well as ways to prevent those errors in the future.

9. **References:** Any information borrowed from another source which is not common knowledge must be cited within the text of the report. All sources of information are to be listed in the reference section of the lab report.

10. **Additional Notes:** Reports will be graded largely on their ability to clearly communicate results and important conclusions to the reader. You must, of course, use proper language and spelling, along with comprehensible logic and appropriate style. You should proof-read and spell-check your report. Neatness and organization will also influence the grade a report receives. Be sure to

follow explicitly the format indicated above. Do not copy material without citing the source. This includes lab manuals, text books, your neighbor, old labs, etc. Plagiarism, of any degree, will not be accepted. Two sample lab reports, common sections of a lab report and instructors marking guide are provided at end of this manual.

Module 1: Volumetric and gravimetric quantitative analysis
Module description

The practical activities in this module provide students with the basic hands on knowledge in analytical chemistry. It covers wet analytical chemistry techniques. The aim of the practical activities is to familiarize the students with the knowledge required for conducting quantitative analysis. Practical laboratory experiments included in this module are preparation of solutions, standardization techniques, volumetric mixture analysis, precipitation and complexometric titrations, redox titrations and gravimetric quantitative determinations. The methods are employed for the quantitative determination of specific analyte concentrations in various environmental samples.

On successful practices and lab activities, students are expected to:
- Use analytical equipment for quantitative determinations
- Quantitatively analyze the measured results by analytical equipment
- Independently and intelligently evaluate the best analysis method and
- Develop new analytical techniques.

Experiment 1
Preparation of solutions by dissolving solid samples

Introduction

The key step in chemical analysis is preparation of solutions since most quantitative and qualitative analyses are performed by converting the analyte of interest in a solution of appropriate grade and concentration. Solutions of known concentration can be prepared in a number of different ways depending on the nature of the analyte and the concentration required. Among the different techniques of preparation of laboratory solutions are: Weighing out a solid material of known purity, dissolving it in a suitable solvent and diluting to the required volume, weighing out a liquid of known purity, dissolving it in a suitable solvent and diluting to the required volume, diluting a solution previously prepared in the laboratory and diluting a solution from a chemical supplier. From the masses and volumes used during preparation of solutions we can calculate the concentrations of each solution using the relation:

$$concentration = \frac{amount\,of\,solute}{amount\,of\,solution}$$

Common units of concentrations of solutions are molarity, normality, molality, mole-fraction, percent by mass or percent by volume, ppm and ppb.

When making solutions for use in the science lab, controlling the concentration of solutes in the solution is often critical. For example, failure to carefully control the concentration of solutes in solutions used to culture cells can cause the cells die or develop abnormally. Likewise, inappropriate concentrations of solutes in solutions that contain bio-molecules may cause the molecules to denature or aggregate, thus making them inactive. Concentrations of solutes can also affect the speed and types of chemical reactions that take place within a solution.

In this lab, students will examine a method for preparing solutions with specific concentrations of solutes by dissolving solid solutes in a solvent.

Dissolving solid solutes in the solvent

This method must be used when there is no existing stock solution. In this case, you dissolve the solid solute (usually granules or powder) in the solvent (usually distilled H_2O). To make a solution in this way, you must know how much of the solute to use and how much solvent to mix with it. When preparing a solution by dissolving a solid solute in the solvent, use the following formula to calculate the weight of solute needed:

Weight of solute (g) = formula weight of solute (g/mole) x molarity (mole/L) x final volume (L)

Purpose

The objective of this experiment is to learn and practice methods of preparing laboratory solutions from solid samples.

Chemicals and apparatus

o   solid NaCl, distilled water

o   beakers, volumetric flasks

o   graduated cylinders, pipette

o   electronic balance

Procedure

1. Determine the mass of solute you need (e.g.NaCl), and weigh it out with an electronic balance.

2. Place the solute in an appropriate volumetric measuring device. For example, if you need 100 mL of solution, a 150 mL graduated cylinder would be an appropriate choice.

3. Carefully add water to your measuring device until the bottom of the meniscus reaches the level of the final volume of your solution.

4. Seal the top of the measuring device so no solution can leak out.

5. Gently invert the measuring device several times until all of the solute is dissolved.

Post Lab

The formula weight of $KMnO_4$ is 158.04 g/mole. How much potassium permanganate should you weigh out in order to prepare one liter of the following $KMnO_4$ solutions? a) 2 M b) 0.75 M

Experiment 2
Preparation of solutions by diluting an existing stock solution

Introduction

In order to maximize all available storage space most solutions are stored in a concentrated form known as stock solution. These solutions are then diluted to the required strength as and when required for the individual experiments.

When preparing solutions, it is often easier to dilute an existing stock solution than it is to weigh out a solid solute and then dissolve it in the solvent. This is especially true if we want to prepare several solutions that each have a different concentration of the same solute.

When making parallel dilutions, we use the following formula to calculate the amount of stock solution needed for each dilution: $C_1V_1 = C_2V_2$, where

C_1 is the concentration of the starting (stock) solution

V_1 is the volume of starting (stock) solution needed to make the dilution (this is our unknown)

C_2 is the desired concentration of you final (dilute) solution

V_2 is the desired volume of you final (dilute) solution

Purpose

The objectives of this experiment are to learn and practice methods of preparing laboratory solutions

Chemicals and apparatus

o Beakers, volumetric flasks

o graduated cylinders, pipette

o Conc. HCl (12 M)

Procedure

1. Calculate the required amount of stock solution (V_1) (e.g. Conc. HCl)

2. Pour V_1 mL of the stock solution into a graduated cylinder and then add enough distilled H_2O to bring the final volume up to V_2.

3. Alternatively, you can pour the required amount of stock solution into a container, and then calculate and add the amount of distilled H_2O needed to give you a final volume V_2.

Post Lab

You want to prepare one liter of 2 M potassium permanganate ($KMnO_4$) solution. How many moles of potassium permanganate should you dissolve in one liter of solution?

Experiment 3
Standardizing Sodiumhydroxide (NaOH) and Hydrochloric acid (HCl) solutions

Introduction

In some instances, the concentrations of some laboratory solutions may not be accurately known due to unexpected physical and/or chemical changes such as hygroscopy, effervescence, evaporation, dissociation and presence of impurities in the solution. It is therefore necessary to determine the concentrations of diluted solutions by titrating against primary standard solutions. The accurate determination of concentration of a solution whose concentration is only approximately known by titrating against a primary standard solution is called standardization. A primary standard solution is one whose composition is constant with time and has a well known concentration.

During titration, it is therefore important to know the exact concentration of the titrant (the solution in the burette which will be added to the unknown) in order to determine the concentration of the solution being tested. In this experiment, we will standardize the ~0.1 M NaOH solution (the titrand) with potassium hydrogen phthalate (KHP, $KC_8H_4O_4H$) using phenolphthalein as the indicator. KHP is a weak acid and reacts with base as:

$$C_8H_4O_4H^- + OH^- \rightarrow C_8H_4O_4^{2-} + H_2O$$

We will also standardize 0.1 M HCl solution (the titrant) with sodium carbonate (Na_2CO_3) using phenolphthalein as the indicator. Na_2CO_3 is a base and reacts with the strong acid HCl in this way: $2HCl + Na_2CO_3 \rightarrow NaCl + H_2O + CO_2$

Purpose

This experiment is aimed at learning how to prepare secondary standard solutions.

Chemicals and apparatus:
- Dry KHP (~1 g), Na_2CO_3 (~0.5 g)
- NaOH solution, HCl solution, Phenplphthalein indicator
- Burette, Erlenmeyer flasks, beakers

Procedure

Part A: Standardizing NaOH:

1. Weigh 1 g of dried KHP (MW = 204.23 g/mol) into an Erlenmeyer flask and dissolve in   60 mL of distilled water. Record the amount of KHP and water used.

2. Add 4 drops of indicator into the flask and titrate with NaOH to the first permanent appearance of pink. Near the endpoint, add the NaOH dropwise to determine the total volume most accurately.

3. Do not discard the remaining NaOH. You will use this for the rest of these experiments.

Data and calculations

Data Table: Part A	
Mass of KHP, g	
Moles of KHP, mol	
Volume of water used, mL	
Conc. of KHP, mol/L	
Moles of NaOH, mol/L	
Volume of NaOH used, mL	
Conc. of NaOH, mol/L	

Part B: standardizing HCl

1. Weigh 0.5 g Na_2CO_3 into an Erlenmeyer flask and dissolve it in 50 mL of boiled, cooled distilled water. Record the exact amount of Na_2CO_3 used. (The water is boiled to expel CO_2 from the solution.)

2. Add 4 drops of phenolphthalein to the solution and record the color.

3. Titrate with the HCl until just before the endpoint (when the solution is very light pink) and then gently boil the solution to expel the CO_2 from solution that has been produced during the reaction.

4. Cool the solution to room temperature and then wash the sides of the flask with a small amount of H_2O to get the entire sample back into solution.

5. Record the color of the solution and the volume of HCl used.

Data and calculations

Data Table: Part B	
Mass of Na_2CO_3, g	
Moles of Na_2CO_3, mol	
Conc. of Na_2CO_3, mol/L	
Moles of HCl, mol/L	
Volume of HCl used, mL	
Conc. of HCl, mol/L	

Experiment 4
Titrimetric analysis of a mixture of Na₂CO₃ and NaOH using two indicators

Introduction

The analysis of a mixture containing Na_2CO_3 and NaOH requires two consecutive steps: the phenolphthalein end point and the methyl orange end point. During titration of the mixture with HCl to the phenolphthalein end point, all hydroxide and half of the carbonate react with the acid. Let the volume of HCl used to this point be V_1. Titration of the mixture with HCl to the methyl orange end point consumes all the hydroxide and all the carbonate. We let the volume of HCl used to reach the methyl orange end point be V_2. Therefore,

Volume of HCl required to neutralize half of the carbonate = $V_2 - V_1$.

Volume of HCl required to neutralize all carbonate = $2(V_2 - V_1)$

Volume of HCl required to neutralize NaOH = $V_2 - 2(V_2 - V_1)$

Purposes

This experiment has the purpose of determining the amounts of NaOH and Na_2CO_3 in the same solution. It also helps practicing using double indicators for analysis of mixtures of acids or bases

Chemicals and apparatus

o   mixture of NaOH and Na_2CO_3

o   HCl (0.1N)

o   phenolphthalein and methyl orange

o   beakers, flasks, burette, ring stand

Procedure

1. With a pipette, transfer 20 ml of the mixture to a conical flask, and add one or two drops of phenolphthalein.

2. Titrate with the 0.1 N HCl till the solution becomes colorless.

3. Repeat the experiment twice or three times and tabulate your results.

4. Repeat the experiment with methyl orange until the color of the solution is changed to faint red.

5. Repeat the experiment twice or three times and tabulate your results.

6. Calculate the concentration of the sodium hydroxide and the sodium carbonate in the mixture.

Data and calculations

Data Table	
V_1, mL	
V_2, mL	
Conc. NaOH, N	
Conc. NaOH, g/L	
Conc. Na_2CO_3, N	
Conc. Na_2CO_3, g/L	

Calculations

1. Normality of Na_2CO_3 x volume of mixture = Normality of HCl x Volume of HCl to titrate all carbonate

2. Normality of NaOH x volume of mixture = Normality of HCl x Volume of HCl to titrate NaOH

Experiment 5
Precipitation Titration: Determination of Chloride by the Mohr Method

Introduction

Chlorides are widely distributed in water as salts of sodium, calcium, potassium and magnesium. Chlorides associated with sodium exert a salty taste to water when its concentration exceeds 250 mg/L in water supplies intended for public consumption. The amount of chloride in aqueous solutions is determined by precipitation titration. Titration is a process by which the concentration of an unknown substance in solution is determined by adding measured amounts of a standard solution that reacts with the unknown. Then the concentration of the unknown can be calculated using the stoichiometry of the reaction and the number of moles of standard solution needed to reach the end point.

Precipitation titrations are based upon reactions that yield ionic compounds of limited solubility. The most important precipitating reagent is silver nitrate. Titrimetric methods based upon silver nitrate are sometimes termed argentometric methods. Potassium chromate can serve as an end point indicator for the argentometric determination of chloride, bromide and cyanide ions by reacting with silver ions to form a brick-red silver chromate precipitate in the equivalence point region. The Mohr method uses chromate ions as an indicator in the titration of

15

chloride ions with a silver nitrate standard solution. After all the chloride has been precipitated as white silver chloride, the first excess of titrant results in the formation of a silver chromate precipitate, which signals the end point. The reactions are:

$$Ag^+ + Cl^- \rightarrow AgCl(s)$$
$$2Ag^+ + CrO_4^{2-} \rightarrow Ag_2 CrO_4 \ (s)$$

By knowing the stoichiometry and moles consumed at the end point, the amount of chloride in an unknown sample can be determined.

Purposes

To standardize $AgNO_3$ solution

To determine the percent of chloride in a solid sample

Chemicals and apparatuses

o $NaCl$, $CaCO_3$, $NaHCO_3$, K_2CrO_4

o $AgNO_3$, buret, transfer pipette, pipette pump

o   Erlenmeyer flasks, desiccators, volumetric flask, amber bottle

o   Graduated cylinder, Wash bottle

Procedure

Preparation of 5% K_2CrO_4 (indicator)

1 g of K_2CrO_4 was dissolved in 20 mL of distilled water.

Preparation of standard $AgNO_3$ solution

1.10.0 g of $AgNO_3$ was weighed out, transferred to a 500 mL volumetric flask and made up to volume with distilled water.

2. The resulting solution was approximately 0.1 M. This solution was standardized against $NaCl$.

3. Reagent-grade $NaCl$ was dried overnight and cooled to room temperature.

4.0.50 g portions of $NaCl$ were weighed into Erlenmeyer flasks and dissolved in about 150 mL of distilled water.

5. In order to adjust the pH of the solutions, small quantities of $NaHCO_3$ were added until effervescence ceased.

6. About 2 mL of K_2CrO_4 was added and the solution was titrated to the first permanent appearance of red Ag_2CrO_4.

Determination of Cl^- in solid sample

1. The unknown was dried at 110 °C for 1 hour and cooled in a desiccator.

2. Individual samples were weighed into 250-mL Erlenmeyer flasks and dissolved in about 100 mL of distilled water.

3. Small quantities of $NaHCO_3$ were added until effervescence ceased.

4. About 2 mL of K_2CrO_4 was introduced and the solution was titrated to the first permanent appearance of red Ag_2CrO_4.

5. An indicator blank was determined by suspending a small amount of chloride free $CaCO_3$ in 100 mL of distilled water containing 2 mL of K_2CrO_4.

Data and calculations: A. Standardization of $AgNO_3$

Data Table : standardization of $AgNO_3$			
Trial	Mass of sample, g NaCl	Volume of $AgNO_3$ used, mL	Conc. of $AgNO_3$, M
Blank	-		
1	0.50		
2	0.75		
3	0.50		

B. Determination of Chloride in Unknown

Data Table: determination of chloride			
Trial	weight of unknown, g	volume of $AgNO_3$,mL	% of Cl^- in unkown
1	0.40		
2	0.50		
3	0.40		

Post lab

In your report,

1. Show all calculations to determine the concentration of $AgNO_3$

2. Explain why the titration is run in neutral conditions

3. Indicate all possible sources of errors in the experiment

Experiment 6
Determination of Hardness of Water by EDTA Titration
Introduction

The hardness of water is defined in terms of its content of calcium and magnesium ions. One of the factors that establish the quality of a water supply is its degree of hardness. Since an analysis does not distinguish between Ca^{2+} and Mg^{2+}, and since most hardness is caused by carbonate deposits in the earth, hardness is usually reported as total parts per million of calcium carbonate by weight. A water supply with a hardness of 100 parts per million would contain the equivalent of 100 grams of $CaCO_3$ in 1 million grams of water or 0.1 gram in one liter of water. In the days when soap was more commonly used for washing clothes, and when people bathed in tubs instead of using showers, water hardness was more often directly observed than it is now, since Ca^{2+} and Mg^{2+} form insoluble salts with soaps and make a scum that sticks to clothes or to the bath tub. Detergents have the distinct advantage of being effective in hard water, and this is really what allowed them to displace soaps for laundry purposes.

The ions involved in water hardness, i.e. Ca^{2+}(aq) and Mg^{2+}(aq), can be determined by titration with a chelating agent, ethylenediaminetetraacetic acid (EDTA), usually in the form of disodium salt (H_2Y^{2-}). The titration reaction is:

$$Ca^{2+}(aq) + H_2Y^{2-}(aq) \rightarrow CaY^{2-}(aq) + 2H^+(aq)$$

Eriochrome Black T is commonly used as indicator for the above titration. At pH 10, Ca^{2+}(aq) ion first complexes with the indicator as $CaIn^+$(aq) which is wine red. As the stronger ligand EDTA is added, the $CaIn^+$(aq) complex is replaced by the CaY^{2-}(aq) complex which is blue. The end point of titration is indicated by a sharp colour change from wine red to blue. Titration using Eriochrome Black T as indicator determines total hardness due to Ca^{2+}(aq) and Mg^{2+}(aq) ions. Hardness due to Ca^{2+}(aq) ion is determined by a separate titration at a higher pH, by adding NaOH solution to precipitate $Mg(OH)_2$(s), using hydroxynaphthol blue as indicator.

Purposes

To determine the concentrations of Ca^{2+}(aq) and Mg^{2+}(aq) ions in a sample of bottled mineral water.

To compare experimental results with the concentrations of the metal ions claimed by WHO

Chemicals and apparatus

- Ammoniumchloride, Ammoniumhydroxide, Erichrome black T
- EDTA, Magnesiumsulfate, hydroxynaphthol blue
- NaOH solution, burette, pipette
- conical flask, graduated cylinder, wash bottle
- beaker, standard flask

Procedure

Part A: Determination of total hardness

1. Pipette 50 cm^3 mineral water (tap water) into a conical flask.

2. Add 2 cm^3 buffer solution followed by 3 drops of Eriochrome Black T indicator solution.

3. Titrate with 0.01 M EDTA until the solution turns from wine red to sky blue with no hint of red (save the solution for colour comparison).

4. Repeat the titration to obtain two concordant results.

Part B: Determination of concentration of Ca^{2+}(aq) ions

1. Pipette 50 cm^3 of mineral water (tap water) into a conical flask.

2. Add 30 drops of 50% w/v NaOH solution, swirl the solution and wait for a couple of minutes to completely precipitate the magnesium ions as Mg(OH)$_2$(s).

3. Add a pinch of hydroxynaphthol blue (exact amount to be decided by the intensity of the resulting coloured solution) and titrate with 0.01 M EDTA until it changes to sky blue.

4. Repeat the titration to obtain two concordant results.

Results and calculation

Part A: Determination of total hardness

	Trial 1	Trial 2	Trial 3
Final burette reading/cm^3			
Initial burette reading/cm^3			
Volume used/cm^3			
Average Volume =			

Part B: Determination of concentration of Ca^{2+}(aq) ions

	Trial 1	Trial 2	Trial 3
Final burette reading/cm^3			
Initial burette reading/cm^3			
Volume used/cm^3			
Average Volume =			

Calculation:

Total hardness as mg/L of $CaCO_3$ equivalent = $(V_{edta} \times N_{edta} \times 50 \times 1000)/ V_{sample}$ = a

Ca^{2+} content as mg/L of $CaCO_3$ equivalent = $(V_{edta} \times N_{edta} \times 20 \times 1000)/ V_{sample}$ = b

Mg^{2+} content as mg/L of $CaCO_3$ equivalent = a-b

Experiment 7
Standardization of sodiumthiosulphate solution with standard $K_2Cr_2O_7$ solution

Introduction

As the thiosulphate is a secondary standard solution, it has to be standardized by titrating against a primary standard dichromate solution iodometrically using starch indicator. Iodometry refers to the titration of iodine liberated in a quantitative redox reaction by a standard solution of a reducing agent like sodium thiosulphate. In iodometry, an aqueous solution of potassium iodide is added to an acidic solution. Oxidants such as dichromate can oxidize iodide quantitatively and rapidly, liberating an equivalent amount of iodine, which is then determined by titration with sodium thiosulphate solution, which has been standardized. Oxidants having higher standard reduction potential than iodine can quantitatively and rapidly oxidize iodide to iodine usually in acid medium.

$$Cr_2O_7^{2-} + 14\,H^+ \rightleftharpoons 2Cr^{3+} + 7H_2O \qquad E^0 = 1.33\ V$$
$$I_2 + 2\,e^- \rightleftharpoons 2\,I^- \qquad E^0 = 0.54\ V$$
$$S_4O_6^{2-} + 2e^- \rightleftharpoons 2S_2O_3^{2-} \qquad E^0 = 0.08V$$

In iodometry a species is titrated with an iodide solution and then the released iodine is titrated with thiosulphate whereas in iodimetry, a species is directly titrated with an iodine solution. Therefore, iodometry is an indirect method and iodimetry is a direct method. Iodometry can be used to quantify oxidizing agents, whereas iodimetry can be used to quantify reducing agents

20

Purpose

This experiment is aimed at getting acquainted with preparation of a secondary standard solution (sodiumthiosulphate) to use for iodometric titrations.

Chemicals and Apparatus

- o   Sodium thiosulphate, Potassium dichromate (0.1 N),
- o   Sodium hydrogen carbonate, Conc. HCl, KI(20%),
- o   Starch (1.0%), burette, pipette, conical flask,
- o   Measuring cylinder, dropper, distilled water

Procedure

1. Pipette out 20 mL of standard $K_2Cr_2O_7$ solution in a 250 mL conical flask.

2. Add 2 mL of 20% KI followed by 4 gm of $NaHCO_3$ and 10 mL of conc. HCl.

3. Swirl the mixture well, cover with watch glass and stand it for 5 min. in dark place to ensure complete liberation of iodine.

4. Wash down the watch glass and sides of the flask with distilled water and dilute it to 200 mL.

5. Then immediately titrate the solution against thiosulphate until the brown color fades to straw yellow.

6. Add 4 mL of 1% freshly prepared starch solution shake the solution to obtain a deep blue color.

7. The titration is continued with continuous shaking until the blue color just disappears leaving a bright green solution.

8. Note the end point and repeat the process till three concurrent readings are obtained.

9. Calculate the concentration of thiosulphate solution using $N_1V_1 = N_2V_2$.

Results and Discussion

Data Table: Measured Volumes of standard K_2CrO_7 solution			
Trial No	Initial burette reading, mL	Final burette reading, mL	Volume of K_2CrO_7
1			
2			
3			
Average volume			

Experiment 8
Determination of the weight percent of iron by redox titration

Introduction

        Iron is one of the most frequently encountered elements in environmental analyses. Its accurate determination is, therefore, of great practical importance. A volumetric determination of iron by redox titration consists of these steps: preparation of the sample, addition of special reagents to aid in detection of the endpoint and to ensure that the proper reaction occurs during the subsequent titration and titration of iron (II) to iron (III) with a suitable oxidizing agent. Major components of an iron sample are often iron (II) and iron (III) compounds that are insoluble in water. These can be dissolved in hot concentrated hydrochloric acid. Silica, which is present in most natural iron samples, does not dissolve in hot concentrated hydrochloric acid and may be observed as a white solid material floating on the surface of the solution; silica will not interfere with the iron determination.

        Potassium permanganate, $KMnO_4$, is a relatively inexpensive strong oxidizing agent. Permanganate, MnO_4^-, has an intense dark purple color. During reduction of purple permanganate ion to the colorless Mn^{2+} ion, the solution will turn from dark purple to a faint pink color at the equivalence point. No additional indicator is needed for this titration. The reduction of permanganate requires strong acidic conditions.

        In this experiment, permanganate will be standardized by reduction with oxalate ions, $C_2O_4^{2-}$, in acidic conditions. Oxalate ion reacts very slowly at room temperature so the solutions are titrated hot to make the procedure practical. The unbalanced redox reaction is shown below.

$$MnO_4^- + C_2O_4^{2-} \rightarrow Mn^{2+} + CO_2 \text{ (in acidic solution)}$$

        In part I of this experiment, a potassium permanganate solution will be standardized against a sample of potassium oxalate. Once the exact normality (eq/L) of the permanganate solution is determined, it can be used as a standard oxidizing solution. In part II of this experiment, the standard permanganate solution will be used to find the concentration of iron (II) in a ferrous solution (g/L). The unbalanced redox reaction is shown below.

$$MnO_4^- + Fe^{2+} \rightarrow Mn^{2+} + Fe^{3+} \text{ (acidic solution)}$$

Phosphoric acid will be used to ensure that the ferric product, Fe^{3+} remains in its colorless form.

Purpose

The objective of this experiment is to determine the percentage of iron in an unknown by photometric redox titration using a solution of potassium permanganate. The first step is to determine the precise concentration of the $KMnO_4$ solution, since it is impossible to obtain $KMnO_4$ in a reproducibly pure state.

Part I: Preparation and standardization of 0.1 N $KMnO_4$ Solutions.

Equipment and Reagents

- o $KMnO_4$ (solid), weighing paper, burette, 500 mL Florence flask
- o $K_2C_2O_4$ H_2O, ring stand, rubber stopper, analytical balance
- o   Burette clamp, hot plate or Bunsen burner,   250 mL Erlenmeyer flask, 6 N H_2S

Pre-lab Questions

1. 12.5 grams of $Na_2C_2O_4$ are dissolved to make 250.0 mL of solution. How many moles of $Na_2C_2O_4$ are present in each 1.00 mL of this solution?

2. If 1.00 mL of the solution (above) is titrated with a $KMnO_4$ solution in order to determine the concentration of the $KMnO_4$. The end point occurs after the addition of 200 uL. What is the molar concentration of the $KMnO_4$ solution?

3. Assuming that 1.25 grams of an unknown ferrous salt is dissolved in 250 mL of solution. If a 1.00 mL sample of this solution requires 193.0 uL of the same $KMnO_4$ solution, what is the % of iron in the unknown?

Procedure

1. On a centigram balance, weigh about 1.0 g $KMnO_4$ crystals on a piece of weighing paper. Add the crystals to a 500 mL Florence Flask.

2. Add about 350 mL of distilled water to the flask.

3. Heat the solution with occasional swirling to dissolve the $KMnO_4$ crystals. Do not boil the solution. This may take about 30 minutes.

4. Allow the solution to cool and stopper. You will need this solution for procedures that follow.

5. On weighing paper, weigh about 0.2 – 0.3 g of hydrated potassium oxalate, $K_2C_2O_2.H_2O$, on the analytical balance. Record the exact mass. Transfer the sample to a 250 mL Erlenmeyer flask.

6. Rinse and fill the burette with the $KMnO_4$ solution.

7. Add 50 mL of distilled water and 20 mL of 6 N H_2SO_4 to the oxalate sample in the Erlenmeyer flask. Swirl to dissolve the solids.

8. Heat the acidified oxalate solution to about 85 °C. Do not boil the solution.

9. Record the initial burette reading. Because the $KMnO_4$ solution is strongly colored, the top of the meniscus may be read instead of the bottom.

10. Titrate the hot oxalate solution with the $KMnO_4$ solution until the appearance of a faint pink color.

11. Record the final burette reading and calculate the volume of $KMnO_4$ used in the titration.

12. Discard the titration mixture down the drain and repeat the titration with a new sample of oxalate for a total of 2 trials.

13. An oxalic acid solution may be used to wash the burette and the titration flask if a brown stain remains in the glassware.

Results and Calculations

Data Table: Volumes of $KMnO_4$ used for titration of Oxalate Solution		
Trial No	Initial volume reading, mL	Final volume reading, mL
1		
2		
3		
Average volume		

1. Using the half-reaction method, write a balanced redox equation for the reaction of permanganate with oxalate in an acidic solution.

2. Calculate the equivalent weight of the oxalate reducing agent from the molar mass of the oxalate sample and the equivalence of electrons lost by the reducing agent in the oxidation half-reaction.

$$equivalent\ wt. = \frac{molar\ mass}{moles\ of\ electrons}$$

3. Use the sample mass and the equivalent weight to calculate the number of equivalents of oxalate in each sample.

$$equivalents\ of\ reducing\ agent = \frac{actual\ mass}{equivalent\ mass}$$

At the equivalence point, the equivalence of the reducing agent is equal to the equivalence of the oxidizing agent.

$$equivalents\ of\ reducing\ agent = equivalents\ of\ oxidizing\ agent$$
$$N_{red}V_{red} = N_{ox}V_{ox}$$

4. Calculate the normality of the $KMnO_4$ solution from the equivalence of the oxidizing agent and the volume used in the titration.

5. Calculate the average normality of the permanganate solution.

Part II: Determination of the Mass of Iron in a Ferrous Solution.

Equipment and Reagents

- Unknown Fe^{2+} solution $KMnO_4$ solution   Burette
- Clamp   250 mL Erlenmeyer flask   25 mL pipette
- Ring stand   6 N H_3PO_4 pipette bulb

Procedure

1. Pipet a 25 mL sample of the unknown Fe^{2+} solution into a 250 mL Erlenmeyer flask.

2. Add 50 mL of distilled water and 12 mL of 6 N H_3PO_4 into the flask.

3. Fill a burette with the standard $KMnO_4$ solution and record the initial burette reading.

4. Titrate the sample with the standard $KMnO_4$ to a faint pink end-point and record the final burette reading. Calculate the volume of $KMnO_4$ used.

5. Discard the ferric solution down the drain and repeat the titration with a new sample of the ferric solution for a total of 2 trials.

6. When finished with all trials, discard the purple permanganate solution in the appropriate waste container in the fume hood.

7. Oxalic acid may be used to remove any brown stains left on the glassware.

Results and Calculations

1. Using the half-reaction method, balance the redox reaction of permanganate with iron (II) in acid.

2. Calculate the equivalence of $KMnO_4$ titrated.

$$Equivalents\ of\ oxidizing\ agent = N_{ox}V_{ox}$$

At the equivalence point, the equivalence of the oxidizing agent is equal to the equivalence of the reducing agent, Fe^{2+}.

3. Determine the normality of the ferrous reducing agent

$$N_{Fe} = \frac{equivalents\ of\ Fe}{Volume\ of\ solution}$$

4. Calculate the molarity (mol/L) of the ferrous solution: $MFe = NFe/n$

5. Calculate the mass concentration (g/L) of iron in the unknown solution by multiplying the molar mass of iron by the molarity of the ferrous solution: $mass\ concentration = mol/L \times 56g/mol = g/L$

6. Calculate the average mass concentrations for the ferrous unknown solution.

Experiment 9

Gravimetric Determination of Calcium as Calcium Oxalate

Introduction

Calcium is an essential nutrient for the body. It is involved in the normal function of nerves and muscles including the heart. Calcium ions in the blood are also necessary for blood clotting. In order to prevent blood from clotting, it is necessary that the calcium ions be precipitated out by adding some other substance. Blood samples used in the laboratory often have potassium oxalate added to them (causing calcium oxalate to precipitate out). Since oxalates are poisonous, during transfusions sodium citrate is added which causes solid calcium citrate to form.

A deficiency in the amount of calcium ion in the blood stream will cause calcium from bones to dissolve to replace it. Calcium is normally obtained through consuming milk and milk products. However, this can be supplemented by taking calcium tablets. Commercial tablets primarily consist of calcium carbonate (chalk). The solid calcium carbonate tablets are dissolved by stomach acid (hydrochloric acid, HCl) freeing the calcium ion to go into solution:

$$CaCO_3(s) + 2\ HCl(aq) \rightarrow CaCl_2(aq) + CO_2(g) + H_2O$$

The concentration of calcium in a sample can be determined by gravimetric analysis. In this experiment an unknown Ca^{2+} containing sample will be analyzed by precipitating the Ca^{2+} using oxalate ($C_2O_4^{2-}$). In the presence of basic oxalate solution, Ca^{2+} forms an insoluble precipitate.

$$Ca^{2+}(aq) + C_2O_4^{2-}(aq) \rightarrow CaC_2O_4.H_2O(s) \qquad\qquad K_{sp} = 1.7 \times 10^{-9}$$

The resulting precipitate is, however, soluble in the presence of acidic solution because the oxalate anion is a weak base. Large relatively pure crystals that are easily filtered will be obtained if the precipitation is carried out slowly. This can be done by dissolving Ca^{2+} and $C_2O_4^{2-}$ in acidic solution and gradually raising the pH by thermal decomposition of urea:

$$H_2N-CO-NH_2 + 3H_2O + heat \rightarrow CO_2 + 2NH_4^+ + 2 OH^-$$

Chemicals and Apparatus

o   Calcium carbonate tablets, 0.1 M HCl, 250 – 400 mL
o   beakers, methyl red indicator, ammonium oxalate solution,
o   stirring glass rod, 15 g of solid urea, filter funnel, suction generator, ice

Purpose

This experiment has the objective of determining the amount of calcium in a calcium carbonate tablet sample by precipitating the ion as calcium oxalate.

Pre-lab Question

1. Why shouldn't you handle your sintered glass funnels with your hands (other than not wanting to burn yourself)?

Procedure

1. Dry three medium-porosity sintered-glass funnels for 1 – 2 h at 110 ^{0}C. Cool them in a desiccator for 30 mins and weigh them. Repeat this procedure with 30-min heating periods until successive weighings agree to within 0.3 mg. Use paper, paper towel or tongs, ***not your fingers***, to handle the funnels. NOTE: if you placed your funnels in the oven during the last lab period you will probably only need to reweigh once to achieve agreement.

2. Transfer exactly 12.5 mL of unknown to each of three 150 – 200 mL beakers and dilute each with 37.5 mL of 0.1 M HCl.

3. To each of the above solutions add 3 drops of methyl red indicator solution. This indicator is red below pH 4.8 and yellow above pH 6.0.

4. Add 12.5 mL of ammonium oxalate solution to each beaker while stirring with a glass rod. Remove the glass rod and rinse it into the beaker. Add 10 g of solid urea to each sample, cover with a watch glass, and boil gently for 30 minutes until the indicator turns yellow.

5. Filter each hot solution through a weighed funnel, using suction generated by the aspirator. Add 3 mL of ice-cold water to the beaker, and use a rubber policeman to help transfer the remaining solid to the funnel. Repeat this procedure with small portions of ice-cold water until all of the precipitate has been transferred. Finally, use

two 10 mL portions of ice-cold water to rinse each beaker, and pour the washings over the precipitate.

6. Dry the precipitate, first with aspirator suction for 1 min, and then in an oven at $110\ ^0C$ for $1 - 2$ hour. Bring each filter to constant mass. The product is somewhat hygroscopic, so only one filter at a time should be removed from the dessicator, and weights should be done rapidly.

7. Calculate the average molarity of Ca^{2+} in your unknown solution.

Post-lab Questions

1. You are performing what is ideally a very accurate determination of Ca^{2+} by weighing the amount of a precipitate. This determination assumes that your precipitate is pure. What are known problems that can affect the purity of a precipitate?

2. How would you determine whether or not your precipitate was pure?

Experiment 10

Determination of concentration of acetic acid in commercial vinegar solutions

Introduction

Vinegar is a solution of acetic acid (CH_3COOH). The concentration of the vinegar is usually given on the label of the bottle in percent by weight or "percent acidity." It is recommended by some food and drug administration agencies that the product called simply "vinegar" is made from apples and contains not less than 4 g of acetic acid in 100 mL of vinegar. One way to produce cheap vinegar is to keep the concentration at this allowable minimum. However, vinegar may gradually lose strength on the shelf so the manufacturer may wish to make the product stronger than necessary in order to guarantee a good shelf life. Vinegars that are not made from apples are available, including malt vinegar, made from barley and corn, wine vinegar and rice vinegar. Whatever the source, acetic acid is the "sour", or acid, ingredient.

In this experiment, we will determine the acetic acid concentration of vinegar collected from local markets by titration with 1 M sodium hydroxide solution. Different vinegars may have different subtle flavoring agents, nevertheless, the vinegar acts as a source of water-soluble acid in food preparation. Other acid sources that are sometimes used are lemons and sour milk. Any vinegar sample may be used, but colorless vinegar is preferred because it gives less interference with the observation of the indicator endpoint color change. As the concentration of the vinegar solution is much higher than the concentration of your standardized sodium hydroxide solution, the original vinegar solution is diluted ten times prior to the titration. This dilution factor must be taken into account when calculating the concentration of the original vinegar solution.

Purpose

This experiment is aimed at finding the concentration of acetic acid in some commercial vinegar samples collected from local markets using titration method. The result is compared with allowed levels of the acid in vinegar.

Chemicals and apparatus

- o   500 mL plastic bottle, 100 mL beaker, 25.00 ml volumetric pipette
- o   250.0 mL volumetric flask, 250 mL Erlenmeyer flask, 50 mL Burette
- o   Vinegar solution, deionized water, phenolphthalein indicator solution
- o   standardized sodium hydroxide solution

PROCEDURE

1. Measure exactly 12.5 mL of vinegar into a clean 250 mL volumetric flask   .

2. Dilute the vinegar with deionized water to the mark on the volumetric flask   .

3. Stopper the flask and mix the solution well. (Invert the solution slowly for at least 10 times to completely mix the contents).

4. Transfer the dilute vinegar solution to a clean and dry 500 mL plastic bottle and label it with the contents

5. Immediately wash your volumetric flask with plenty of tap water and several portions of deionized water. Let the flask dry at room temperature

6. Pour about 25 mL of the dilute vinegar solution in 100 mL beaker

7. Rinse your 25 mL volumetric pipette several times with portions of diluted vinegar from your beaker.

 8. Carefully pipet 25 mL of diluted vinegar solution into the 250 mL Erlenmeyer flask

 9. Add about 50 mL of deionized water to the Erlenmeyer flask

10. Add 2 drops of phenolphthalein indicator solution and swirl the flask to thoroughly mix the solution

11. Rinse your 50 mL burette several times with a few milliliters of your standardized sodium hydroxide solution

12. Fill the burette with your standardized sodium hydroxide solution. Record the volume or the buret to the nearest 0.01 mL

13. Titrate the acid sample to a faint pink end point. Record the final volume of the burette to the nearest 0.01 mL

14. Repeat the titration procedure described above for at least two more three trials.

Results and calculation

Data Table: Determination concentration of acetic acid by titration with NaOH			
Quantity determined	Trial 1	Trial 2	Trial 3
Final burette reading (mL)			
Initial burette reading (mL)			
Volume of NaOH used (mL)			
Moles NaOH used (mol)			
Moles of $HC_2H_3O_2$ used			
Volume of diluted vinegar			
Molarity of diluted vinegar (M)			
Average molarity of diluted vinegar (M)			
Average molarity of original vinegar (M)			

Post lab

Assuming the density of vinegar to be 1.5002 g/mL, calculate the % (w/w) of acetic acid in vinegar

Experiment 11
Determination of iodate in iodized salt by redox titration
Introduction

Iodine is an essential trace element for human nutrition. The safe dietary intake of iodine as recommended by the World Health Organization (WHO) is 100 μg day^{-1} for infants and 150 μg day^{-1} for adults. Iodine is required by the thyroid gland for the synthesis of growth hormones. The storehouse of iodine in the human body is the thyroid gland. Inadequate intake of iodine leads to iodine deficiency symptoms and disorders like goiter, extreme fatigue, mental retardation, and depression which are collectively called as iodine deficiency disorders (IDDs). Milk, vegetables, fruits, cereals, eggs, meat, spinach, and sea foods are the natural dietary sources of iodine. However, natural sources of iodine may not satisfy its requirement by the body as iodine from these sources may not be in a bioavailable form and also the concentration of iodine may be below the recommended level.

Adequate intake of iodine can be achieved by consumption of iodized salt. Iodization of salt is done by addition of iodate to commercial salt products due to its good stability and bioavailability. Thus, determination of iodate in commercial salt is of considerable importance as the amount of iodate in the salt samples may vary with environmental conditions, the nature of transport, packing conditions, and cooking methods.

In this Experiment we will determine the amount of iodate (IO_3^-) in iodized salt by reacting the iodate with iodide (I^-), under acid conditions to produce iodine according to the following reaction.

$$IO_3^- + 5\,I^- + 6\,H^+ \rightarrow 3\,I_2 + 3\,H_2O$$

The resulting iodine is then titrated with thiosulfate as follows.

$$I_2 + 2\,S2O_3^{2-} \rightarrow 2\,I^- + S_4O_6^{2-}$$

The amount of iodine, I_2, titrated with thiosulfate is determined stoichiometrically by using the equation

$$N_1V_1 = N_2V_2,\ \text{Where 1 refers to iodine and 2 for thiosulfate.}$$

Chemicals and Apparatus

o   iodized salt, 0.002 M sodium thiosulfate solution

o   1 M hydrochloric acid, 0.5% starch indicator solution

o   0.6 M potassium iodide solution (10 g solid KI made up to 100 mL with distilled water)

o   250 mL volumetric flask, 50 mL pipette (or 20 and 10 mL pipettes)

o   250 mL conical flasks, 10 mL measuring cylinder

o   burette, distilled water, ring stand

Purpose

This experiment has the objective of determining the level of iodine in some commercial table salt samples using a method of reduction-oxidation titration.

Procedure

1. Preparation of 0.002 M sodium thiosulfate solution

Accurately weigh about 1.25 g of solid sodium thiosulfate ($Na_2S_2O_3 \cdot 5H_2O$) and dissolve in 50 mL of distilled water in a volumetric flask. This gives a 0.1 M thiosulfate solution. Then use a pipette to transfer 10 mL of this solution to a 500 mL volumetric flask and dilute by adding distilled water up to the mark; we will use this diluted thiosulfate solution in our titrations. Notice that the concentration of thiosulfate may be determined more accurately by titration with a standard solution of iodate or potassium permanganate

2. Preparation of 0.5% starch indicator solution

Weigh 0.125 g of soluble starch into a 100 mL conical flask or beaker and add 25 mL of distilled water. Heat the solution with stirring at 79°C for 5 minutes. Be careful not to exceed the stated temperature. Allow solution to cool to room temperature.

3. Determination of iodine in iodized table salt

1. Accurately measure 25 g of iodized salt into a 125 mL volumetric flask and add distilled water up to the mark. Make sure all the salt dissolves.

2. Use a pipette to transfer a 50 mL aliquot of salt solution into a 125 mL conical flask. Then add 5 mL of 1 M hydrochloric acid and 5 mL of 0.6 M potassium iodide solution. The solution will turn a yellow/brown colour as iodine is produced.

3. Titrate the solution with your 0.002 M sodium thiosulfate solution until the yellow/brown colour of iodine becomes very pale. Then add 1 mL of starch indicator solution, which will produce a dark blue-black coloured complex with iodine and continue your titration until this colour completely disappears.

4. Repeat the titration with further aliquots of your salt solution until concordant results (titres agreeing within 0.1 mL) are obtained.

Important remark

According to the specified limits for iodate in iodised salt, the volume of 0.002 M sodium thiosulfate required in the above titration should lie between 5.9 mL and 15.4 mL. Therefore, a "rougher" method for quickly determining whether or not your salt sample conforms to these limits is to prepare the sample solution as above (adding hydrochloric acid and potassium iodide as described), but instead of titrating the solution simply add 1 mL of starch indicator, followed by 5.9 mL of thiosulfate solution, the blue-black colour of starch-iodine should persist. But when a further 9.5 mL of thiosulfate solution is added the colour should disappear.

Results and discussion

Data Table: determination iodine in iodized table salt by titration	
Quantity determined	Result
average volume of thiosulfate solution, mL	
moles of thiosulfate, mol	
moles of iodate in salt solution, mol	
molarity of iodate in salt solution, M	
grams of iodate in salt solution, g	
iodate content in mg/kg of salt	
iodine content in mg/kg of salt	

Experiment 12
Determination of the active ingredients in commercial anti-acid tablets

Introduction

Antacids are compounds that act as weak bases to neutralize stomach acid, hydrochloric acid (~0.016M). Most antacids such as calcium carbonate and magnesium hydroxide are water insoluble and thus difficult to analyze by direct titration. Because of the insolubility and any carbon dioxide production, antacids are analyzed by an indirect method or what is known as a **back titration method**. The antacid is allowed to react with a known amount of hydrochloric acid and then the excess hydrochloric acid is titrated with sodium hydroxide. The amount of acid reacted or neutralized by the antacid is determined by difference:

moles of HCl neutralized = initial moles of HCl - moles of excess HCl

If carbon dioxide is produced in the reaction of an antacid, the carbon dioxide must be removed before titration. The carbon dioxide gas can be removed by boiling, which shifts the equilibrium in the reaction below to the left.

$$CO_2 + H_2O \rightleftharpoons HCO_3^- (aq) + H+ (aq)$$

33

If CO_2 is left dissolved, the H^+ produced would be titrated by sodium hydroxide.

Various commercial anti-acids are claimed to be most effective in relieving stomach indigestion. Regardless of their effectiveness, anti-acid tablets have a common purpose of neutralizing excess stomach acid.

The most common weak base ingredients in anti-acid tablets are: aluminum hydroxide, ($Al(OH)_3$), magnedium hydroxide ($Mg(OH)_2$), calcium carbonate ($CaCO_3$), magnesium carbonate ($MgCO_3$), sodium bicarbonate ($NaHCO_3$) and potassium bicarbonate ($KHCO_3$)

To decrease the possibility of the stomach getting too basic from the anti-acids, buffers are added as part of the formulations of some commercial anti-acids. The most common faster relief anti-acids that buffer excess acid in the stomach are those containing $NaHCO_3$ and/or $CaCO_3$. The HCO_3^-/CO_3^{2-} buffer system is established in the stomach with these anti-acids. Excess HCl is used to destroy the buffer action in the anti-acid solution which is later neutralized with a standard solution of NaOH.

Purpose

In this experiment, we will determine the neutralizing effectiveness of the active ingredient in some commercial anti-acids.

Procedure

1. Into a pre-weighed massing boat add one crushed ant-acid tablet. Determine the mass of the tablet added to the nearest 0.1 mg. Quantitatively transfer the tablet to an Erlenmeyer flask.
2. To the crushed ant-acid, add exactly 12.5 mL of standard HCl and approximately 12.5 mL of distilled water. Gently swirl to dissolve the antacid. Heat the solution to boiling and allow it to boil for 5 minutes to ensure complete loss CO_2. Allow the flask to cool before adding indicator and starting the titration.
3. Add 2-3 drops of phenolphthalein to the cooled flask and titrate the excess HCl with standard NaOH. Save the first titration of each brand as a color comparison for the endpoint. Repeat the preparation and titration on 2 additional tablets of the same brand.
4. Analyze a different brand of antacid tablet following the procedure above.

5. For all titrations complete the data and results table given below.

6. For each brand of antacid, calculate the percentage of antacid compound in a tablet and percentage error using the formulas below:

$$\% \ anti \ acid = \frac{mass \ of \ anti \ acid \ by \ titration}{original \ mass \ of \ tablet} \ x \ 100 \ \%$$

$$\% \ error = \frac{mass \ of \ anti \ acid \ by \ titration \ - mass \ of \ anti \ acid \ by \ label}{mass \ of \ anti \ acid \ by \ label} \ x \ 100 \ \%$$

Data Table				
Brand------------------ Anti-acid compound--------------- amount on label---------				
Measured parameter	Trial 1	Trial 2	Trial 3	Average
Mass of massing boat, g				
Mass of massing boat and anti-acid, g				
Volume of HCl added, mL				
Volume of NaOH used, mL				
Calculations and Results				
Mass of anti-acid tablet, g				
Initial moles of HCl				
Moles of HCl in excess				
Moles of HCl neutralized				
Mole of anti-acid in tablet				
Mass of anti-acid in tablet				
% of ant-acid in tablet				
% error				

Post lab

Discuss how the results for % antacid compare to the amount specified on the label. Using the average of the three trials, determine which brand is more effective and why. Suggest some possible sources of error in this determination and how they might affect results.

Module II: Instrumental methods of chemical analysis
Module description

This module introduces the students to some of the methods used for instrumental quantitative chemical analysis, with reference to their relevance in the analysis of real samples. The module extends and builds on the scope to include instrumental methods used for quantitative chemical analysis covering the principal chromatographic, electrochemical and spectrophotometric techniques. They will cover the underlying principles of a range of different techniques and carry out experimental analyses to illustrate the methods, all the while being aware of good laboratory practice when working in the laboratory. They will be expected to present and analyze data in an appropriate manner, including the appropriate statistical treatment of both acquired and presented data. In addition, learners will develop key skills in how to design and put together a poster presentation based on a key technique relevant to the analysis of environmental samples and be able to critically assess the methodology and suggest alternative methods of analysis for the applications discussed.

On successful completion of this module, students will:

- Demonstrate and show familiarity of the terminology associated with analytical chemistry,
- Be able to critically assess and produce a protocol for an appropriate forensic analysis including an appropriate sampling strategy
- Be able to provide a correct statistical analysis of all empirical data
- carry out valid analyses using a variety of analytical techniques, in an accurate and reliable manner
- Produce reports and be able to present analytical data, including the experimental results to a prescribed pattern and including an appropriate data analysis using the principles of valid analytical measurements

Experiment 1
Extraction of Benzoic acid
Introduction

The principle of liquid–liquid extraction (LLE) is that a sample is distributed or partitioned between two immiscible liquids or phases in which the

36

compound and matrix have different solubilities. Normally, one phase is aqueous (often the denser or heavier phase) and the other phase is an organic solvent (the lighter phase). The basis of the extraction process is that the more polar hydrophilic compounds prefer the aqueous (polar) phase and the more non-polar hydrophobic compounds prefer the organic solvent. Alternatively, if the target organic compounds are to be analyzed by gas chromatography they are best isolated in an organic solvent. The compounds in the organic solvent (for GC) can be analyzed directly or pre-concentrated further using, for example, solvent evaporation, while compounds in the aqueous phase (for HPLC) can be analyzed directly or pre-concentrated further using, for example, solid phase extraction. The main advantages of LLE are its wide applicability, availability of high purity organic solvents and the use of low-cost apparatus (e.g. a separating funnel).

Extraction is frequently employed in the laboratory to isolate one or more components from a mixture. Unlike re-crystallization and distillation, it does not yield a pure product; thus, the former techniques may be required to purify a product isolated by extraction. In the technical sense extraction is based on the principle of the **equilibrium distribution** of a substance (solute) between two immiscible phases, one of which is usually a solvent. Extraction is a convenient method for separating an organic substance from a mixture, such as an aqueous reaction mixture or a steam distillate. The extraction solvent is usually a volatile organic liquid that can be removed by evaporation after the desired component has been extracted.

As it is discussed above, the technique of extraction is based on the fact that if a substance is insoluble to some extent in two immiscible liquids, it can be transferred from one liquid to the other by shaking it together with the two liquids. For example, acetanilide is partly soluble in both water and ethyl ether. If a solution of acetanilide in water is shaken with a portion of ethyl ether (which is immiscible with water), some of the acetanilide will be transferred to the ether layer. The ether layer, being less dense than water, separates out above the water layer and can be removed and replaced with another portion of ether. When this in turn is shaken with the aqueous solution, more acetanilide passes into the new ether layer. This new layer can be removed and combined with the first. By repeating this process enough times, virtually all of the acetanilide can be transferred from the water to the ether. Extraction is accomplished by shaking the solution in a **separatory funnel** with a second solvent that is immiscible with the one in which the compound is dissolved, but dissolves the compound more readily. Two liquid layers are formed, and the layer that has most of the desired product in it can be separated from the other. Sometimes not the entire product is extracted in a single operation and the process must be repeated once or twice more to assure a clean separation. It has been found that when two immiscible solvents are shaken together, the solute distributes itself between them in a ratio roughly proportional to its solubility in each. The ratio of the concentration of the solute in each solvent at equilibrium is a constant called the **distribution ratio (d) or partition coefficient (K_d).** The larger the value of K_d, the more solute will be transferred to the ether with each extraction, and the fewer portions of ether will be required for essentially complete removal of the solute.

$$K_d = C(org)/C(aq) = (W_{org}/V_{org})/(W_{aq}/V_{aq}) \qquad (1)$$

Where C(org) = concentration of solute in organic phase,

   C(aq) = concentration of solute in aqueous phase,

   W(org) = weight of solute in organic phase,

   V(org) = volme of organic solvent,

   W(aq) = weight of solute in aqueous phase,

   V(aq) = volume of aqueous phase.

The percentage of solute extracted into the organic layer, % E, is given as

% E= (weight of solute extracted/total weight) x 100% $\qquad$ (2)

For multiple extractions, % E = $[1- (V_{aq}/(V_{aq} + K_dV_o))^n]$ $\qquad$ (3)

   where n is the number of exactions.

Criteria for selecting an extracting solvent

1) It should be insoluble or slightly soluble with the solvent of the solution being extracted.

2) It should have a favorable distribution coefficient for the substance being extracted and an unfavorable distribution.

3) It should be able to be easily removed from the extracted substance after the extraction. Since the removal is often by distillation, the solvent should therefore have a reasonably low boiling point.

4) It should be chemically inert to the extracted substance, other components in the mixture, and the solvent to the solution being extracted.

5) It should be reasonably safe to work with and relatively inexpensive.

Purposes

To learn practicing the extraction (separation) of Benzoic acid in organic solvent
To determine the distribution coefficient, K_d

Chemicals and apparatus

- Benzoic acid solution   0.1 M NaOH solution   methylene dichloride
- ring & ring stand   burette and burette clamp   spatula
- separatory funnel   Erlemeyer flask   graduated cylinders
- beakers   funnel

Procedure

A. Determination of concentration of benzoic acid

1. Use a 10 mL graduated cylinder to measure 10.0 ml of Benzoic acid solution and transfer the solution to a 125 mL Erlenmeyer flask.
2. Add 2-3 drops of phenolphthalein and titrate to the end point (light pink) with a standardized ($\approx$ 0.1M or 0.02 M) sodium hydroxide solution.
3. Record in report form the number of milliliters of base required to neutralize this volume of acid solution. Calculate the molarity of the Benzoic acid solution in two more trials.

B. Determination of concentration of benzoic acid extracted in dichloromethane.

1. Use a 50 mL graduated cylinder to measure out a second 50.0 ml volume of benzoic acid solution and transfer it to your separatory funnel.
2. Add 10 ml of methylene dichloride, CH_2Cl_2, to the funnel and extract according to the procedure outlined by the instructor or laboratory technician.
3. Separate the *bottom* layer (organic phase) in a beaker and collect the *top* layer (aqueous) into an Erlenmeyer flask and add 2-3 drops of indicator.
4. Record the volume of the sodium hydroxide solution in the burette and titrate to the phenolphthalein end point (light pink).
5. Again record the number of milliliters of base required

Results and Calculation

Table1: Volume of base required to neutralize 10.0 ml of the benzoic acid solution

Trial No	1	2	3	Average
Volume				

39

Calculations

Molarity of benzoic acid (BA) = $(M_{NaOH} \times V_{NaOH})/V_{BA}$ = ----------------M (a)

Volume of base required to neutralize 50 ml of benzoic acid solution after a single extraction with 10.0 ml of CH_2Cl_2 = ------------------------ ml (b)

Moles of benzoic acid neutralized = a x b/1000 = --------------- mole (c)

Grams of benzoic acid neutralized= c x (M_W of benzoic acid) = --------g (d)

Grams of benzoic acid originally present in 50 ml before extracting with $CH_2 Cl_2$ ------g (e)

Grams of benzoic acid extracted in CH_2Cl_2 = (e – d) = --------------g (f)

Percent of benzoic acid in extracted in CH_2Cl_2 = (f/e) x 100 % = ------------% (g)

Distribution coefficient Kd = C (CH_2Cl_2)/C (H_2O) = (f/10 ml)/ (d/50ml) =------- (h)

Experiment 2
Paper Chromatography

Introduction

Paper chromatography is one form of planar chromatography in which a piece of filter paper or cellulose is used as the stationary phase and a liquid or a mixture of liquids is used as the mobile phase. Among several forms of paper chromatography, ascending paper chromatography is the most popular where the mobile phase flows upward through the stationary phase.

During mixture analysis using paper chromatography, a tiny drop of the mixture to be separated is placed on the paper near the bottom. A lightly drawn pencil line marks the location of the spot. This location is called the **origin**. The paper is suspended vertically in the mobile phase, a solvent or **eluent**. The eluent could be water or alcohol, or a solvent solution made from several reagents whose proportions are chosen to enhance their ability to "pull" along some substances in the mixture being separated better than others. We want each chemical in our mixture to have different attractions to the solvent so that they will travel at different speeds and be separated. The **origin** must be above the surface of the **eluent**. The eluent rises up the paper by capillary action. When the eluent reaches the origin, the components of the mixture rise at different rates. The container must be covered to prevent evaporation of eluent. The chromatogram must be removed from the eluent *before* the eluent reaches the top of the paper. As the substances in the mixture rise up the paper, they spread out and the spots become larger. For this reason, the original spot should be as small as possible, less as 5 mm in diameter. If too much material is applied to the small spot, the spot may develop a long "tail." If too little material is applied to the spot, the color of the spot may be too faint to see as the spot enlarges while moving up the paper. Trial-and-error and experience help the experimenter obtain both a small spot and one with the proper amount of material.

Substances can be identified by the heights they reach on the completed chromatogram by calculating R_f (rate of flow or retention factor) values. The R_f value is a constant for a given substance under the same experimental conditions. The R_f value may be calculated from the following equation.

$$R_f = d_{solute}/d_{solvent}$$

Where d_{solute} is distance of the center of the sample spot from the origin and $d_{solvent\ is}$ distance of the solvent front from the origin

Figure 1 shows the finished chromatogram of substance A, substance B, and a mixture containing substances A and B. To determine the distance traveled by each component measure the distance from the origin to the center of the migrated spot. If the spot is large with a "tail," measure to the "center of gravity" or densest concentration of the spot.

R_f (substance A) = 4.2 cm/14.8 cm = 0.284, R_f (substance B) = 9.6 cm/14.8cm = 0.649

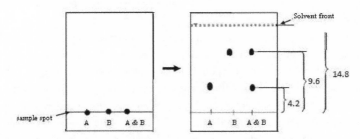

Figure 1: Typical finished chromatogram of two substances

Once the R_f value is known, the substance can sometimes be identified by comparing its R_f value with those reported in the literature. To check the identity of an unknown substance, it is usually necessary to run a chromatogram of a known sample simultaneously with the unknown.

Purposes

To determine the R_f values for some selected ink samples.

To use standardized R_f values to identify the colored components in an unknown mixture.

Chemicals and apparatus

o   Isopropanol used as a solvent, black felt pen ink of different colors,

o   Role of filter paper   developing chamber.

Procedure

1. Prepare two parts isopropanol to one part water as a solvent

2. Obtain a strip of chromatography paper about 2.5 cm wide by 10 cm tall.

3. Along one of the shorter sides, draw a horizontal line in pencil about 1.5 cm from the edge of the strip. This will be the "base line", the starting line where the samples will be spotted. Graphite will not be carried up the chromatography paper. This baseline will be used later in the calculations of R_f values.

42

4. Using the black felt tip pens, apply dots of ink on the baseline

5. With a pencil, label each dot with the identity of the pen from which the ink came.

6. Obtain a paper strip from your instructor that is marked with one of the three pens. This is the unknown ink sample that you must identify.

7. Once your strips are prepared, set up a developing chamber using a 400 ml beaker.

8. Pour some of the developing solvent in the beaker, using only enough solvent to cover the bottom of the beaker.

9. Place the paper strip into the developing chamber so that the edge near the ink spots is submerged in the solvent. *The ink spot must be above the solvent level.* You should avoid allowing the paper strips to touch each other or the sides of the beaker. (Taping the top of the strips to the outside of the beaker may be helpful.) Once you have inserted the paper, cover the chamber with a watch glass and make sure the solvent is progressing up the paper.

10. When the solvent has risen almost to the top of the paper strip, remove the strip from the chamber and immediately mark the level to which the solvent has risen and circle each spot present on the strip with a pencil. (The solvent front and the spots will continue to move for several minutes after removing the strip from the beaker.)

11. Measure the distance each spot traveled and calculate R_f values for each spot. Use the circles you drew around each spot with your pencil and not the actual spot!

Results and calculation

Data Table: Calculation of the Rf values of some dyes or indicators

color	Distance moved by solvent(cm)	Distance moved by solute(cm)	R_f

Post lab

Compare the R_f values that you calculate with the standard values.

Experiment 3
Thin-Layer Chromatography

Introduction

This experiment will introduce the technique of thin-layer chromatography (TLC). TLC is a different type of planar chromatography in which the stationary phase is a thin layer of silica gel (a sand-like substance) coated onto a piece of plastic, glass or aluminum foil. The thin plate provides the solid support for the stationary phase. Note that some TLC stationary phases are activated by heating them prior to use. An appropriate LLC mobile phase solvent or a mixture of solvents is used as the mobile phase. While the solvent and dissolved mixture passes through the stationary phase, components of the mixture adsorb to the surface of the silica gel and are held back. Therefore, some molecules move through the silica gel more quickly than others. In TLC the compounds interact with silica gel based on their polarity. Silica gel is very polar; and polar compounds interact more strongly with the silica gel than non-polar compounds. Thus, polar compounds move more slowly than non-polar compounds. The distance that a compound travels relative to the mobile phase is referred to as its retention factor or R_f value. In order to separate the substances of a mixture, the substances must have different R_f values. By carefully choosing an eluent and a sorbent, it is usually possible to find a combination that will separate the mixture. R_f value for a solute is characteristic of that substance in a given set of experimental conditions or with a specific solvent. Thus, the components of a mixture may be identified by comparing measured Rf values with a standard Rf value for each solute.

In this part of the experiment, students will be given, or asked to determine experimentally, the R_f values for a number of indicators. They will then use thin-layer chromatography to separate and identify some of these substances present in an unknown mixture. TLC can be utilized to separate small quantities of organic compounds.

Thin-layer chromatography is almost identical to paper chromatography. Instead of using paper, the stationary phase is a thin coating of adsorbent material, called the **sorbent**, on a sheet of glass, plastic, or metal foil. As in paper chromatography, the TLC plate is suspended vertically in an eluent and the eluent travels up the plate. TLC offers two advantages over paper chromatography. First, it provides a better separation of the mixture with less spreading of the spots; second, the sorbent may be varied.

Common sorbents include compounds like Silica (SiO_2), very pure, finely ground sand), alumina (Al_2O_3, also used in abrasives, ceramic materials, and dental cement), and cellulose (similar to very pure, finely ground wood fibers).

Purposes

To determine the R_f values for some selected indicators used in chemistry.

To use standardized R_f values to identify the colored components in an unknown mixture.

Chemicals and apparatus

o 2% by volume isobutyl alcohol, 2% by volume concentrated aqueous ammonia

o 96% by volume water, methyl orange, indigo carmine, methyl red indicators

Procedure

1. Obtain a TLC plate from your instructor. Be careful not to bend the plate excessively since the silica adsorbent may flake off. In addition, a TLC plate should only be handled by the edges.

2. Set the plate down on a clean, dry surface, then lightly draw a line across the plate about 1.0 cm from the bottom of the plate with the aid of a PENCIL as shown below.

3. Next make three 2-3 mm lines, spaced about 0.5 cm apart and running perpendicularly through the line across the bottom of the TLC plate. Be sure that the first and last lines are about 0.5 cm from the edge of the plate. These lines mark the points where the samples will be spotted or applied to the TLC plate

4. Pour 2.5 – 3.0 ml of the mixture of isobutyl alcohol + water + ammonia into a 500 ml beaker.

5. Put points of several dyes above the first line onto the strip of the TLC paper.

6. Observe the separation of the mixture and fill the result table.

Results and calculation

Data Table: Calculation of the R_f values of some dyes or indicators

Dye	Distance moved by solvent(cm)	Distance moved by solute(cm)	R_f

Post lab

Compare the R_f values that you calculate with the standard values.

Experiment 4
Conductometric Measurements of strong electrolytes
Introduction

Conductometric analysis involves measurement of the electrolytic conductivity of solutions. The electrolytic conductivity is a sum parameter, resulting from the individual conductivities of the ions in the solution. The conductivity is a function of the concentration, which is linear within certain concentration ranges. The ion mobility, which determines the alternating current conductivity, depends on type of ion (size and hydration), the polarity of the solvent, the temperature and the viscosity. Direct conductometric concentration measurements can be made using in-expensive equipment in salt solutions or pure acids and bases. The main application of the conductometric analysis is the determination of an end point in volumetric analysis. The method is also widely used to check the total electrolyte content of water.

Movement of ions in water can be studied by immersing a pair of electrodes into the liquid and by applying a potential difference between the electrodes. Like metallic conducting materials, electrolyte solutions follow Ohm's law:

$$R = V/I \tag{1}$$

where R is the resistance in ohms(Ω), V is the potential difference in volts, and I is the current in amperes.

The conductance G of a solution in units of Siemens or Ω^{-1} is then defined as reciprocal of the resistance:

$$G = 1/R \tag{2}$$

Conductance of a given liquid sample decreases when the distance between the electrodes increases and increases when the effective area of the electrodes increases. This is shown in the following relation:

$$G = \kappa A/l \tag{3}$$

where κ is the specific conductivity (S m^{-1}), A is the cross-sectional area of the electrodes (m^2); i.e. the effective area available for conducting electrons through the liquid), and l is the distance between the electrodes (m).

The Ratio l/A is an intrinsic parameter of each conductometric container and is called **cell constant, θ**.

$$G = \kappa/\theta \tag{4}$$

Molar conductivity (Λ_m(S m^2 mol^{-1}), the conductance of 1 molar solution of an electrolyte is defined as:

46

$$\Lambda_m = \kappa/c \qquad\qquad (5)$$

where c is the molar concentration of the added electrolyte.

The molar conductivity of an electrolyte would be independent of concentration if κ were proportional to the concentration of the electrolyte. In practice, however, the molar conductivity is found to vary with the concentration (Figure 1). One reason for this variation is that the number of ions in the solution might not be proportional to the concentration of the electrolyte. For example, the concentration of ions in a solution of a weak acid depends on the concentration of the acid in a complicated way, and doubling the concentration of the acid does not double the number of ions. Another issue is that ions interact with each other and tend to slow down each other leading reduced conductivity. In this limit, the molar conductivity depends on square root of electrolyte molar concentration.

In the 19th century Friedrich Kohlrausch discovered the following empirical relation between the molar concentration of a strong electrolyte and the molar conductivity (Kohlrausch's law) at low concentrations:

$$\Lambda_m = \Lambda_m^\circ - k\sqrt{c} \qquad\qquad (6)$$

where k is a non-negative constant depending on the electrolyte and Λ_m° is the limiting molar conductivity (e.g. the molar conductivity in the limit of zero concentration of the electrolyte).

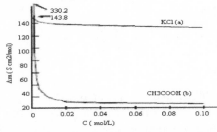

Figure 1: Variation of molar conductivity as a function of molar concentration a) Strong electrolute and b) weak electrolyte

Table of specific conductances of KCl solution with the concentration of 0.1mol/l at various temperatures

$T(^\circ C)$	15	16	17	18	19	20	21	22	23	24	25
$\kappa_{KCl}(Sm^{-1})$	1.05	1.07	1.09	1.12	1.14	1.17	1.19	1.22	1.24	1.26	1.29

Purposes

Determination of cell constant, measurement of conductances of KCl solutions
Calculation of specific conductances and molar conductances

Chemicals and apparatus

- Conductometer, 50 ml volumetric flask50 and 100 ml beakers
- Pipette, 0.M KCl solution, distilled water

Procedure

1. Plug the conductometer in and turn it on.

2. Fill the 50 ml beaker up with a solution of 0.1M KCl. Dip the conductometric container into the beaker. Note that all three platinum fillets are immersed in the solution. Read the value of conductance G_{KCl} and take the conductometric container off the solution. Repeat the measurement of G_{KCl} 2 more times and calculate its average value.

3. Determine the temperature in the laboratory, and find tabulated conductance that corresponds to this temperature in the Table 4. Calculate the cell constant, θ using the average value of G_{KCl} and equation 4.

4. Measure cunductances of 0.05, 0.04, 0.03, 0.02, 0.01M solutions of KCl

5. Pour the KCl solution back to the storage bottle. Carefully rinse and dry conductometric container and the beaker.

Results and calculation

Data table 1: conductance of 0.1 mol/l KCl solution

Trial	1	2	3	Average
G_{KCl}(S)				

Data table 2: conductances of different concentrations of KCl solution

Conc. KCl (M)	0.1	0.05	0.04	0.03	0.02	0.01
G (S)						
κ_{KCl}(S m^{-1})						
Λ_m (S m^2 mol^{-1})						

Post lab

Plot Λ_m vs. $\sqrt{c}$, and calculate the value of k (obtain Λ_m° from literature)

Experiment 5
Conductance measurements of weak electrolytes

Introduction

It is possible to differentiate between strong and weak electrolytes by measuring their electrical conductance. Strong electrolytes follow Kohlrausch's law, whereas weak electrolytes are described by Ostwald's dilution law. The examination of the concentration dependence of the conductivity allows the molar conductivities of infinitely diluted electrolytes to be determined, and facilitates the calculation of the degree of dissociation and the dissociation constants of weak electrolytes.

Since weak electrolytes are only slightly ionized in solution, the number of ions is not proportional to the concentration of the electrolyte but depends on the degree of dissociation (α). The effective molar conductivity can then be approximated in terms of α and the molar conductivity of the fully ionized electrolyte at infinite dilution (Λ_m°):

$$\Lambda_m = \alpha \Lambda_m^\circ \tag{1}$$

Ostwald's dilution law is valid for weak electrolytes. It enables dissociation constants to be calculated:

$$K = (\alpha^2.C)/1 - \alpha = (\Lambda^2_m.C)/(\Lambda_\infty - \Lambda_m)\,\Lambda_\infty \tag{2}$$

The limiting value of the molar conductivity of weak electrolytes at infinite dilution is first reached at extremely low concentrations; therefore, exact measurements in this are no longer possible. Consequently Λ_∞ cannot be obtained by extrapolating curves for weak electrolytes. Equation (3) is derived by transforming Ostwald's law of dilution:

$$1/\Lambda_m = 1/\Lambda_\infty + \Lambda_m\,.C/(\kappa\,.\,\Lambda^2_\infty) \tag{3}$$

From equation (3) it can be seen that a linear relationship exists between the reciprocal of the conductivity and the product of the molar conductivity and the concentration of weak electrolytes. Furthermore, Ostwald's law of dilution shows that the molar conductivity at infinite dilution can be obtained from the line's point of intersection of the line with the ordinate $1/\Lambda_m$ over $C.\Lambda_m$

Purposes

To measure conductances of CH_3COOH solutions

To calculate of specific conductances, molar conductances, dissociation constants

Chemicals and apparatus

- Conductometer   50 ml volumetric flask   50 and 100 ml beakers
- pipet   distilled water   0.1 M CH_3COOH

Procedure

1. Plug the conductometer in and turn it on.

2. Measure the conductance of distilled water to ensure that the beaker and conductometric container are clean. The conductance has to be less than 10 mS.

3. Prepare solution with acetic acid concentrations of 0.5, 0.4, 0.3, 0.2 and 0.1M and pour it into 50 ml beaker. Immerse conductometric container into the solution and measure conductance G.

4. Repeat measurement for each of the diluted solutions as well as for the stock solution. Do not rinse conductometric container, neither the beaker in between particular measurements.

5. Calculate the specific conductance of each solution using the measured values of a conductance and cell constant of conductometric container.

6. Rinse the conductometric container and all the beakers carefully at the end of the experiment.

Results and calculations

Data table: Conductance, G, and specific conductance κ of acetic acid solutions

Conc. (M)	0.5	0.4	0.3	0.2	0.1
G (S)					
κ (S m^{-1})					
Λ_m (Sm^2/mol)					
α					
K_a					

Post lab

1. Plot the values of conductances of acetic acid solution as a function of the concentration.

2. Obtain values of Λ_m° from literature and calculate α and then K_a values for each measurement.

Experiment 6
Acid-base titration using a pH meter

Introduction

In an acid-base titration, the important information to obtain is the **equivalence point**. If there are a given number of moles of acid in the titration flask, the equivalence point is reached when those same numbers of moles of base have been added from the buret. The molarity of the base can then be calculated since the number of moles of base added is the same as the number of moles of acid in the flask, and the volume of the base added is also known. Similarly, if the number of moles of acid in the titration flask is unknown, it can be calculated for the equivalence point if the molarity of the base and the volume of base added are known.

Often the pH of the solution will change dramatically at the equivalence point. An acid-base indicator works by changing color over a given pH range. If an indicator which changes color near the equivalence point is chosen, there is also a dramatic change in the color of the indicator at the equivalence point because the pH changes so rapidly.

In a potentiometric acid-base titration, an indicator is not necessary. A pH meter is used to measure the pH as base is added in small increments (called aliquots) to an acid solution. A graph is then made with pH along the vertical axis and volume of base added along the horizontal axis. From this graph the equivalence point can be determined and the molarity of the base calculated

Purposes

To perform a potentiometric titration of an acidic solution of known molarity

To graph the volume of base added vs the pH and to determine the equivalence point

To calculate the molarity of the basic solution

Chemicals and apparatus

- o   50 mL buret buret clamp   ring stand   magnetic stirrer
- o   250 mL beaker   HCl solution   NaOH Solution (unknown)

Procedure

1. Obtain about 100 mL of 0.15 M HCl in a clean, dry beaker. This beaker should be labeled. Never pour any solution back into this beaker. Once the solution has been poured into the burette, it should be discarded into the sink or waste container.

2. Rinse your buret with distilled water. Then use a small amount of the 0.15 M HCl solution to rinse the   burette. (Pour about 10 ml of the HCl solution into the burette. Let some of it flow through the tip.   Pour the rest of the HCl solution out the top of

the burette, rotating the burette as you pour.) The rinsing solution should be discarded into the sink. Repeat this rinsing procedure twice more. Fill the burette to some point higher than the markings with the HCl solution and then carefully let the HCl   solution out into a waste container until the bottom of the meniscus is on the 0.00 line.

3. Obtain about 100 mL of NaOH solution in another clean, dry beaker. This beaker should also be labeled. Never pour any solution back into this beaker. Once the solution has been poured into the burette, it should be discarded into the sink or waste container. Rinse and fill the other burette as indicated in step 2.

4. Standardize the pH meter using pH 4, pH 7 and pH 10 buffer solutions

5. Place a 250 mL beaker under the buret containing the 0.15 M HCl and let out approximately 20 mL into the beaker. Record the mL of HCl added to the beaker.

6. Carefully drop a magnetic stirring bar into the beaker containing the HCl solution. Set the beaker on the magnetic stirring motor and position the burette containing the NaOH solution and the pH electrode as shown in the diagram. Carefully turn on the stirring motor and make sure that the stirring bar does not hit the electrode. Adjust the stirring speed as directed by your teacher.

7. Set up your data table to include mL of NaOH added and the pH of the solution. You should allow for as many as 50 mL of solution.

8. Measure and record the pH of the solution before any NaOH has been added.

9. Add 1.0 mL of NaOH solution carefully from the burette. Record the pH when it has stabilized. Add another 1.0 mL of NaOH and record the pH. Continue adding NaOH in 1.0 mL increments until you have obtained a pH reading greater than 12.

10. Remove the pH electrode from the solution, rinse it with distilled water, and store it as directed by your teacher.

11. The solutions may be discarded down the sink. Rinse the burettes with distilled water and place it upside down in the buret holder to drain

Results and calculation

Data table: pH of Solution after each addition of 1mL HCl

mL of NaOH added	1	2	3	4	5	6	7	8	9	10	11	12
Measured pH												

Post lab

1. Make a graph of the pH vs mL of NaOH added. The pH should be on the vertical axis and the mL of NaOH should be on the horizontal axis. The graph should be of

such a size that 1 mL is represented by 1 square on the graph and the pH scale is spread out as much as possible.

2. There should be a region on your graph where the slope is very steep. Determine the midpoint of this region. This is the equivalence point. Record the mL of NaOH added at the equivalence point as determined from the graph.

3. Use the relationship: $M_A V_A = M_B V_B$ to determine the molarity of the bas

Experiment 7
Conductometric Titrations

Introduction

Conductance of electrolyte depends upon i) number of free ions, ii) changes on the free ions and iii) mobility of the ions on the substitution of one ion by another of different mobility (speed of ions). So, conductometric method can be used to determine the end point of ionic titrations like i) acidimetric titration, ii) precipitation titration, iii) titration involving the formation of complex ion. When hydrochloric acid solution (HCl) is titrated with sodium hydroxide solution (NaOH), the highly mobile hydrogen ions ($\lambda°_{H+}$ = 350 ohm^{-1} cm^{-1}) are progressively replaced by slower moving sodium ions ($\lambda°_{Na+}$ = 50 ohm^{-1} cm^{-1}) and the conductance of the solution decreases. After the end point, the conductance of the solution rises sharply due to the presence of excess, highly mobile hydroxide ion ($\lambda°_{OH-}$ = 198 ohm^{-1} cm^{-1}). Thus the neutralization of a strong acid by addition of a strong base leads to a minimum conductance at the end points. This is due to the disappearance of H$^+$ ions and their replacement by slower moving Na$^+$ ions of the base followed by the presence of highly mobile OH$^-$ ions after the end point. Therefore the nature of the plot (conductance of the solution versus volume of base added) will be as given below

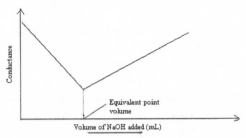

Figure 3: Conductometric titration curve for titration of HCl vs. 0.1M NaOH

The conductivity cell used for this titration should permit stirring by shaking and to which the reagent can be added from burette. A large increase in volume during titration should be avoided.

Purposes

To plot the titration curve and determine the equivalent point of titration

To determine the concentration of a solution of hydrochloric acid

Chemicals and Apparatus

- Conductometer   conductivity cell   beaker
- pipette   burette   conical flask
- Hydrochloric acid   Sodium hydroxide   conductivity water.

Procedure

1. Rinse the conductivity cell a number of times with conductivity water or double distilled water.

2. Pipette out 20 mL of HCl in a beaker and dip the conductivity cell in it, so that the cell should dip completely in solution.

3. Note the temperature of the sample solution and accordingly set the temperature control or keep the cell in a thermostat at room temperature.

4. Add small amount of NaOH solution (few drops) from burette, stir it and measure the conductance after each addition.

5. Take at least five readings beyond the end point.

Results and Calculation

Data table: conductance measurements for Conductometric Titration of HCl with 0.1M NaOH

mL of 0.1M NaOH added	1	2	3	4	5	6	7	8	9	10	11	12
Measured Conductance (S)												

Plot a graph between conductance and volume of titrant (NaOH solution). Two intersecting lines will be obtained (as given in the Figure 1) and the points of intersection of these lines represent the equivalent point.

Let, V_2 be the volume of NaOH at the equivalent point (from graph) and the concentration of acid is C_1 and concentration of NaOH solution is $C_2 = 0.1N$.

Then, $20 \times C_1 = V_2 C_2$

$$C_1 = (V_2 C_2 / 20)$$

Thus, The concentration of the acid is ---------------N

Experiment 8
Absorption spectra of some inorganic compounds

Introduction

Spectrophotometry is a measurement of how much a chemical substance absorbs or transmits electromagnetic radiation and a spectrophotometer is an instrument that can pass light of a single wavelength through a solution and measure the amount that passes through. First, the spectrophotometer must be adjusted to read zero by passing light of the chosen wavelength (500 nm for example) through a blank containing only the solvent. Then, the same wavelength of light is passed through the solution. The percentage of light that passed through the solution relative to the amount that passed through the solvent alone is called the **percent transmittance at 500 nm**. For example, if half as much light passes through a solution as passes through the solvent alone, we would record this as 50% T

$$percent\ transmittance = \frac{\text{amount of light transmitted through the solution}}{amount\ of\ light\ transmitted\ through\ solvent} \; x\ 100$$

On the other hand, the amount of light at 500 nm that is absorbed by the solution is called the **absorbance at 500 nm** and is abbreviated A_{500}. Absorbance is the negative logarithm of the percent transmittance divided by 100. Because logarithms have no units, absorbance has also no units.

$$Absorbance = -log\left(\%\frac{transmittance}{100}\right) = 2 - \log\left(\%\ transmittance\right)$$

Note that absorbance and transmittance are inversely related to each other. In other words, as the absorbance of a particular wavelength increases, the transmittance of that wavelength decreases. It turns out that, within limits, the optical absorbance of a solution at a specified wavelength is **linearly** related to its concentration. This means that if the concentration of a solution increases at a steady rate, then the absorbance will also increase at a steady rate. This is called a **linear relationship** because if we measure the absorbance of several solutions with different concentrations, and then plot a graph with concentration on the x-axis and absorbance on the y-axis, all of our data points should fall on a straight line.

In this experiment, you will be determining the **absorption spectrum** for a $KMnO_4$ solution. An absorption spectrum shows you how much light is absorbed by the solution at various wavelengths.

We will also measure the absorbance of light by the KMnO₄ solution at wavelengths ranging from 480 nm to 580 nm. The wavelength which gives us the highest absorbance is called the **wavelength maximum, λ_{max}**, for the solution.

Purposes

To learn how to operate the spectrophotometer

To plot the absorption spectra of $KMnO_4$ solutions

Chemicals and apparatus

o   Spec-20 spectrophotometer, 2 spec-20 cuvettes

o   volumetric flasks, beakers, test tubes

o $KMnO_4$ solution, label tape

Procedure

Notice that the Spec-20 spectrophotometer has been turned on for at least 20 minutes to warm up and get stable

1. Label a Spec-20 cuvette "B" with a small piece of label tape placed near the top of the cuvette. Fill the cuvette with distilled H_2O. This will be your zero standard or **blank**.

2. Regardless of the type of Spec-20 used, you must perform 3 steps every time you change the wavelength of light used, in this order:

a) adjust to the correct wavelength (using the knob on top of the Spec-20)

b) before placing any cuvette in the instrument, adjust the **% transmittance** to zero (using the **left** knob on the front of the Spec-20)

c) after placing the blank cuvette (labeled "B") in the instrument, adjust the **absorbance** to zero (using the **right** knob on the front of the Spec-20)

3. Set the wavelength of your Spec-20 to 480 nm and adjust the filter if necessary. Remember, the Spec- 20 must be calibrated at 2 points every time you set a new wavelength: the **% transmittance must** be set to zero when the holder is empty, and the **absorbance** must be set to zero when the blank is in the holder.

4. Prepare 0.001M $KMnO_4$, by dilution, to use for determining the absorption spectrum of $KMnO_4$. Choose a sample that has a distinct color, but that is light enough that you can see through it.

5. Draw a data table in your lab notebook to record the absorbance, by the selected solution, of light at the following wavelengths: 480 nm, 500 nm, 520 nm, 540 nm, 560 nm, and 580 nm.

6. Pour the selected solution into a clean, dry cuvette and place the cuvette into the Spec-20 holder.

Notice: Before you place cuvettes into the Spec-20 holder, always wipe the outside of the cuvette with a Kim wipe to remove any fingerprints or dust. Also, make sure the line on the cuvette is lined up with the line on the holder.

7. Read the absorbance at 480 nm and enter the result into the data table in your lab notebook.

Notice: Make sure to measure absorbance and not % transmittance.

8. Remove the cuvette from the Spec-20. Change the wavelength to 500 nm and re-calibrate the Spec-20 spectrophotometer. Remember, you must re-calibrated the Spec-20 at 2 points every time you set a new wavelength: the **% transmittance** must be set to zero when the holder is empty, and the **absorbance** must be set to zero when the blank is in the holder.

9. Place the same solution that you used to measure absorbance at 480 nm back into the Spec-20. Read the absorbance at 500 nm and enter the result into the data table in your lab notebook.

10. Continue in this way until you have measured the absorbance of this same solution at all of the wavelengths listed in the data table of your lab notebook.

11. When you have completed your table for all measurements between 480 nm and 580 nm, examine your data. What is the wavelength maximum for your $KMnO_4$ solution?

12. Set your Spec-20 to the wavelength maximum for $KMnO_4$. In the next experiment, you will use this wavelength to measure the absorbance of **all** your diluted $KMnO_4$ solutions.

Data Table: Absorption spectrum of $KMnO_4$							
Wavelength, λ (nm)	480	500	520	540	560	580	600
Absorbance, A							

Experiment 9
Effects of concentration on absorbance: Beer's law

Introduction

Absorbance measurements can be used in any of several ways in a quantitative analysis. The most straightforward way to determine concentration is to measure absorbance of the sample solution at the wavelength at which the analyte in solution is known to absorb radiation and then use Beer's law which states that the concentration of an analyte in a solution is directly proportional to the absorbance of the solution at a particular wavelength.

Many early scientists studied the passage of light through transparent media such as colored glasses or solutions. Bouguer (1729) discovered that if one piece of glass could halve the intensity of a light beam passing through it, two pieces reduced the intensity not to zero, but to a quarter of the original value. Lambert stated this mathematically: "If a beam of intensity I_o passes through a glass plate and is reduced in intensity to I_o/k, then passage through n plates will result in an intensity transmitted of:

$$I = I_o/ k^n$$

Much later, Beer (1852) showed that this applied to colored solutions, and that if the concentration were halved, then doubling the path length would compensate for the change in the light intensity. That is,

$$I = \frac{I_o}{k^{lC}}$$

where C is the concentration and l is the path length. By taking logarithms, we get to the usual form of the Beer-Lambert law, and define **Absorbance** as

$$A = \log\frac{I_o}{I} = \varepsilon lC$$

in which ε is the **molar absorptivity** of the molecule or ion at the particular wavelength specified.

There are circumstances, however, that some solutions do not obey Beer's law and lead to incorrect calculation of the analyte concentration. It is clearly important to understand and recognize the deviations from Beer's law and the reasons for these deviations. Sources of Errors that cause deviations from Beer's law include:

Instrumental sources: stray light at high absorbance readings, lamp instability at low absorbance readings and wavelength setting error.

Chemical sources: formation of an unwanted complex, dissociation of a desired complex

Operational sources: dirty cells, path-length errors, reflectivity differences, suspended solids, air bubbles and solvent droplets.

Most students tend to assume at first that any instrument will give correct results, particularly if it looks shiny and new. However, this is never a safe assumption; it is always necessary to check an instrument to make sure that it is performing correctly

In this experiment, you will use the wavelength maximum determined in experiment 14 to measure the absorbance of all diluted solutions of $KMnO_4$.

Purpose

This experiment has the objective of Verifying Beer's law from plots of absorbance vs. concentration.

Chemicals and apparatus

- Spec-20 spectrophotometer, 2 spec-20 cuvettes
- volumetric flasks, beakers, test tubes
- $KMnO_4$ solution
- label tape

Procedure

Verification of Beer's law

1. Re-calibrate the Spec-20 with the wavelength set at the wavelength maximum that you determined for $KMnO_4$.
2. Prepare 10, 1, 0.1, 0.01 and 0.005 mM solutions of $KMnO_4$ by diluting the 0.01 M stock solution.
3. Measure the absorbance at the wavelength maximum for each solution that you made using the parallel dilution method. Record the absorbance of each sample in your lab notebook using a clearly labeled data table.
4. Make a plot of Absorbance Vs. concentration and verify if Beer's law is obeyed.

Conc. of $KMnO_4$, mM	10	1	0.1	0.01	0.005
Absorbance at λ_{max}					

Experiment 10
Determination of the concentration of an unknown $KMnO_4$ solution using Beer's law

Introduction

The absorption of light by a substance in a solution can be described mathematically by the Beer-Lambert Law: $A = \mathcal{E}bc$

where: A = absorption at a given wavelength of light,

$\mathcal{E}$ = molar absorptivity, unique to each molecule and varying with wavelength,

b = the path length through the solution that the light has to travel, and

c = the concentration of the solution in moles per liter (molarity).

The more light that is absorbed by a solution, the more deeply colored the solution is in the region of the spectrum that it is not absorbing. So if a solution is deeply colored, we would expect it to have a large absorbance (A) in the wavelength region where it absorbs. That the absorbance (A) would increase with increasing concentration (c) should make intuitive sense: a more concentrated solution absorbs more light because it has more molecules, atoms or ions in it (for a given volume of liquid) to absorb. One drop of food coloring in 5 mL of water will be less deeply colored, and have a corresponding lower value of A, than 4 drops of food coloring in 5 mL of water. The absorbance (A) also increases with increasing path length. This is also somewhat intuitive. If the light passes through 10 cm of the solution, more light will be absorbed (larger A, more deeply colored) than if the light only passes through 1 cm of solution.

The molar absorptivity is unique to whatever the absorbing species is. The larger ε is, the larger the corresponding A will be. Some molecules have enormous molar absorptivities in certain regions of the visible spectrum; these are often used as dyes because a small amount absorbs a great deal of light and thus the substance will appear very deeply colored (in the wavelength region where the dye doesn't absorb). Even if two substances absorb in the same wavelength region and have the same concentration, they may appear very different in terms of the depth of color, due to differences in their molar absorptivities. If the absorption of a particular chromophore is measured in a series of solutions in which its concentrationis varied systematically, a "Beer's Law Plot" can be generated. This plot can act as a "calibration curve" because it can be used to determine the concentration of the chromophore in an unknown solution by measuring the solution's absorbance (A). Note that a proportional relationship exists between the absorbance and the concentration of the chromophore.

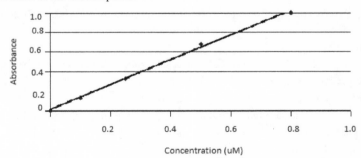

Fig 1: Beer's law plot

60

Once the calibration curve is drawn, the concentration of an unknown sample may be determined by measuring the absorption of the sample at the same wavelength and path length.

Purpose

This experiment has the objective of Verifying Beer's law and determining an unknown concentration of $KMnO_4$ from linear plots of Absorbance vs. Concentration.

Chemicals and apparatus

- o Spec-20 spectrophotometer, 2 spec-20 cuvettes
- o volumetric flasks, beakers, test tubes
- o $KMnO_4$ solution (unknown conc.)
- o label tape

Procedure

1. You will be given a solution with a concentration of $KMnO_4$ that is unknown to you. Measure the absorbance of this solution using the wavelength maximum for $KMnO_4$ that you determined in experiment 2. Record the results in your lab notebook.

2. From plots of Absorbance versus concentration, determine the unknown concentration of $KMnO_4$.

Data Table: Determination of unknown conc. of $KMnO_4$						
Conc. of $KMnO_4$, mM	10	1	0.1	0.01	0.005	Unknown
Absorbance at λ_{max}						

Disposal

Discard the $KMnO_4$ solutions: Make sure you discard all $KMnO_4$ solutions in the waste containers provided; do not dump them into the sink.

Clean the Spec-20 cuvettes: Spec-20 cuvettes are not ordinary test tubes. They are expensive, and great care must be taken to avoid scratching them. Scratches interfere with the passage of light through the tube and that can lead to inaccurate results. Rinse them only---do not use test tube brushes on cuvettes. Rinse the cuvettes thoroughly with tap water and then with distilled H_2O.

Post lab:

Ujulu, a medical technologist in a small rural hospital, uses the Spec 20 for analysis of iron in the blood. Although the ferric ion, Fe^{3+}, is light yellow, it is difficult to detect colorimetrically in dilute solution. Therefore, Ujulu must add several reagents to blood serum, one of which reduces the iron(III) to iron(II), and another which is ferrozine, a compound which complexes the Fe^{2+} ion to form a colored species which, like the red food coloring, can be measured colorimetrically. He obtains the following data:

Conc., µg/dL	0.000	25.00	75.00	125.00	175.00	225.00
Absorbance	0.000	0.144	0.444	0.744	1.08	1.40

1. If the absorbance of blood sample from one patient is 1.17µg/dL, calculate the concentration of iron in the sample.

2. Make a Beer's Law plot of the data for the table above.

3. If the normal serum iron is 40-155 µg/dL, what do you expect the doctor to tell the patient?

Experiment 11
Determination of the composition of a mixture of potassium dichromate and potassium permanganate spectrophotometrically

Introduction

The absorbances of two solutes in a homogeneous solution are additive, provided there is no reaction between the two solutes. Hence we may write

$$A_T(\lambda_1) = A_1(\lambda_1) + A_2(\lambda_1) \qquad (1)$$

$$A_T(\lambda_2) = A_1(\lambda_2) + A_2(\lambda_2) \qquad (2)$$

Where A_1 and A_2 are the measured absorbances at the two wavelengths λ_1 and λ_2 respectively. The subscripts 1 and 2 refer to the two different substances, and the subscripts λ_1 and λ_2 refer to the different wavelengths. The wavelengths are selected to coincide with the absorption maxima of the two solutes; the absorption spectra of the two solutes should not overlap appreciably, so that substance 1 absorbs strongly at wavelength λ_1 and weakly at wavelength λ_2, and substance 2 absorbs strongly at λ_2 and weakly at λ_1. Now, $A = \varepsilon C b$, where ε is the molar absorption coefficient at any particular wavelength, C is the concentration (mol L^{-1}) and b is the thickness, or length, of the absorbing solution (cm). If we set $b = 1$ cm then

$$A(\lambda_1) = \varepsilon_1(\lambda_1)C_1 + \varepsilon_2(\lambda_1)C_2 \qquad (3)$$

$$A\lambda_2 = \varepsilon_1(\lambda_2)C_1 + \varepsilon_2(\lambda_2)C_2 \qquad (4)$$

Solving these equations simultaneously, we obtain

$$C_1 = [\varepsilon_2(\lambda_2)A(\lambda_1) - \varepsilon_2(\lambda_1)A(\lambda_2)] / [\varepsilon_1(\lambda_1)\varepsilon_2(\lambda_2) - \varepsilon_2(\lambda_1)\varepsilon_1(\lambda_2)] \qquad (5)$$
$$C_2 = [\varepsilon_1(\lambda_1)A(\lambda_2) - \varepsilon_1(\lambda_2)A(\lambda_1)] / [\varepsilon_1(\lambda_1)\varepsilon_2(\lambda_2) - \varepsilon_2(\lambda_1)\varepsilon_1(\lambda_2)] \qquad (6)$$

The values of molar absorption coefficients $\varepsilon 1$ and $\varepsilon 2$ can be deduced from measurements of the absorbances of pure solutions of substances 1 and 2. By measuring absorbance of the mixture at wavelengths λ_1 & λ_2, the concentrations of the two components can be calculated. In this experiment, we are concerned with the simultaneous spectrophotometric determination of $K_2Cr_2O_7$ and $KMnO_4$ in a solution

Purpose
The objective of this experiment is to gain hands-on experience with the determination of composition of a binary mixture using UV-VIS spectroscopic method.

Chemicals and Apparatus
0.001 M, 0.0005 M and 0.00025 M solutions of potassium dichromate (in 1 M sulphuric acid and 0.7 M phosphoric acid)
0.001 M, 0.0005 M and 0.00025M solutions of potassium permanganate (in 1 M sulphuric acid and 0.7 M phosphoric acid)

- o   Spec-20 spectrophotometer, 2 spec-20 cuvettes
- o   volumetric flasks, beakers, test tubes
- o   label tape

Procedure
1. Prepare a series of solutions: Potassium dichromate 0.001 M, 0.0005 M and 0.00025 M in mixed solvent of sulphuric acid (1 M) and phosphoric acid (0.7 M).

2. Prepare a series of solutions: Potassium permanganate 0.001M, 0.0005 M and 0.00025 M in sulphuric acid (1M) and phosphoric acid (0.7 M).

3. Measure the absorbance A for each of the three solutions of potassium dichromate and also each of the three solutions of potassium permanganate at both 440 nm and 545 nm by taking 1ml solution each. Calculate ε in each case by using A= εCb and record the mean values for dichromate (2) and permanganate (1) at the two wavelengths.

4. Mix potassium dichromate (0.001 M) and potassium permanganate (0.0005 M) in the following amounts shown in the data table in 100 ml beakers. In each case, total volume of solution should be 25 ml. To each of these solutions add 0.5 ml of concentrated sulphuric acid 5. Measure the absorbance of each of the mixtures at 440 nm and 545nm.

Calculate the absorbance of the mixtures from

$$A_{440} = \varepsilon_{440}C_{Cr} + \varepsilon_{440}C_{Mn}$$
$$A_{545} = \varepsilon_{545}C_{Cr} + \varepsilon_{545}C_{Mn}$$

5. Record the absorbance of the unknown solution (supplied) at 545nm and 440 nm. Calculate the concentrations of permanganate and dichromate in this solution.

Results and discussion

Data Table 1: Preparation of Mixture of $K_2Cr_2O_7$ and $KMnO_4$ solutions			
Volume of $K_2Cr_2O_7$, mL	Volume of $KMnO_4$, mL	Absorbance, measured	Absorbance, calculated
25	0		
20	5		
15	10		
12.5	12.5		
10	15		
5	20		
0	25		

Data Table 2: Absorbances of $K_2Cr_2O_7$ at 440 and 545 nm				
Conc., mM		1	0.5	0.25
Absorbance	440 nm			
	545 nm			
εb	440 nm			
	545 nm			

Data Table 3: Absorbances of $KMnO_4$ at 440 and 545 nm				
Conc., mM		1	0.5	0.25
Absorbance	440 nm			
	545 nm			
εb	440 nm			
	545 nm			

Experiment 12
Effect of chromophores on absorption spectra of Compounds

Introduction

The energy of radiation being absorbed during excitation of electrons from ground state to excited state primarily depends on the nuclei that hold the electrons together in a bond. The group of atoms containing electrons responsible for the absorption is called chromophore. Most of the simple un-conjugated chromophores give rise to high energy transitions of little use. Chromophores, for example, dienes, nitriles, carbonyls, or carboxyl groups often contain π-bonds. Other types of chromophores are transition metal complexes and their ions. Molecules without chromophores, such as water, alkanes or alcohols, should be ideal solvents for UV-visible spectroscopy because they hardly show any absorbance themselves. The following experiment shows that small variations in the structure of molecules can lead to significant differences in the resulting absorbance spectra.

Reagents and Equipment
- o   acetone, acetaldehyde, 2-propanol, distilled water
- o   three 20-ml volumetric flasks, 0.1-ml pipette or syringe
- o   disposable glass pipettes (minimum 4), 10-mm path length quartz cell

Purpose

In this experiment, we will observe differences in the spectra of compounds having different chromophore groups.

Procedure

1. Prepare the following solutions:

    a) 0.1 mL acetone in 20 ml distilled water

    b) 0.2 mL acetaldehyde in 20 ml distilled water

    c) 0.1 mL 2-propanol in 20 ml distilled water

2. Measure a reference on distilled water.

3. Measure the spectra of the acetone, acetaldehyde and 2-propanol solutions in the range from 200 to 350 nm.

Results and discussion

1. Complete the following Table

Compound	λ_{max}	Absorbance
Acetone		
Acetaldehyde		
2-Propanol		

2. Plot the absorption spectra of these compounds

3. Which of these solvents would be best as a solvent for UV-visible analyses?

Experiment 13

Effect of changes in pH on absorption spectra of compounds

Introduction

The absorbance spectrum of a compound is related to its molecular/electronic structure. Changes of the environmental conditions can cause changes in the molecular/electronic structure. A change of the pH value, for example, has an influence on chemical equilibria and can thus change the absorbance spectrum of a solution. The wavelength of the absorbance maximum for bromothymol blue, for example, is shifted to shorter wavelengths (blue shift) with increasing pH value. This color change arises from a shift in the equilibrium of non-dissociated and dissociated dye molecules: H-Indicator $\rightarrow H^+ +$ Indicator$^-$

Wavelengths at which the absorbance does not change are called isosbestic points. Isosbestic points can occur, if the absorbing species of the equilibrium have the same extinction coefficient at a certain wavelength. In this case the absorbance is independent of the position of the equilibrium and depends only on the total amount of the compounds. Isosbestic points are useful for measurements in un-buffered solutions or solutions with unknown pH values because the absorbance reading is independent of the pH as shown below (Figure 1).

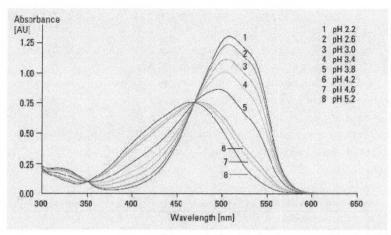

Source: Agilent Technologies, workbook, 2000

Figure 1: Measured absorbance spectra of methylene orange at different pH values

In this experiment we will observe the absorption spectrum of potassium dichromate in acidic and basic media. The potassium chromate/dichromate equilibrium is markedly pH sensitive. The equilibrium in acid solution forms mainly dichromate ions and the one in basic solution mainly forms chromate ions:

$$Cr_2O_7^{2-} + H_2O \rightarrow 2CrO_4^{2-} + 2H^+$$

Reagents and Equipment

o potassium dichromate ($K_2Cr_2O_7$), potassium hydroxide solution 0.1N (KOH)

o hydrochloric acid solution 0.1N (HCl,)distilled water

o 100 mL volumetric flask, two 10 mL volumetric flasks, 10 mL pipette, 1mL pipette

o disposable glass pipettes (minimum 3), 10 mm path length quartz cell

Purpose

This experiment is aimed at observing the effects of changes in pH on absorption spectra of compounds.

Procedure

1 Prepare a stock solution of 6 mg potassium dichromate in 100 ml distilled water.

2 Prepare the sample solutions:

a) Mix 9 ml of the stock solution with 1 ml potassium hydroxide solution 0.1N.

b) Mix 9 ml of the stock solution with 1 ml hydrochloric acid solution 0.1N.

3 Measure a reference on distilled water.

4 Measure the spectrum of the potassium "dichromate" solution in diluted potassium hydroxide solution in the range from 230 to 500 nm.

5 Measure the spectrum of the potassium "dichromate" solution in diluted hydrochloric acid solution in the range from 230 to 500 nm.

Results and discussion

1. Determine the wavelengths of the absorbance maxima of potassium dichromate dissolved in acid and basic solution (2 each) and enter the values in the following table.

Potassium dichromate in	$\lambda_{max(1)}$	$\lambda_{max(2)}$
Acidic solution		
Basic solution		

2. Plot the absorption spectra of the sample in both acidic and basic media on the same graph.

Experiment 14

The Effect of Solvents on UV-visible Spectra of compounds

Introduction

　　　　The UV-Vis spectra are usually measured in very dilute solutions and the most important criterion in the choice of solvent is that the solvent must be transparent within the wavelength range being examined. The table below lists some common solvents with their lower wavelength cut off limits. Below these limits, the solvents show excessive absorbance and should not be used to determine UV spectrum of a sample.

　　　　Highly pure, non-polar solvents such as saturated hydrocarbons do not interact with solute molecules either in the ground or excited state and the absorption spectrum of a compound in these solvents is similar to the one in a pure gaseous state. However, polar solvents such as water, alcohols etc. may stabilize or destabilize the molecular orbitals of a molecule either in the ground state or in excited state and the spectrum of a compound in these solvents may significantly vary from the one recorded in a hydrocarbon solvent.

Solvent	Cut off wavelength(nm)
Acetonitrile	190
Water	191
Cyclohexane	195
Hexane	201
Methanol	203
95% ethanol	304
1,4-dioxane	215
Ether	215
Dichloromethane	220
Chloroform	237
Carbon tetrachloride	257
Benzene	280

Reagents and Equipment
- o benzophenone (mesityl oxide can be used alternatively)
- o Four or five of these solutions: ethanol (CH_3-CH_2-OH), cyclohexane, n-hexane ($CH_3(CH_2)_4CH_3$), acetonitrile (CH_3-CN), methylene chloride (CH_2Cl_2)
- o five 25 mL volumetric flasks, five 100-ml volumetric flasks, 1mL pipette
- o disposable glass pipettes (minimum 5), 1cm path length quartz cell

Purpose

The objective of this experiment is to observe the relationship between solvent polarity and wavelength of absorption maxima.

Procedure

1. Prepare five solutions of about 25 mg benzophenone in 25 mL of the given solvents.

2. Dilute to one-hundredth of the initial concentrations.

3. Measure a reference on the solvents used.

4. Measure the absorbance spectrum of the sample in the range from 300 to 400 nm.

Results and discussion

1. Record the wavelengths of the absorbance maxima of benzophenone for the various solvents in the table below.

Solvent	Dielectric constant	$\lambda_{max (nm)}$
n-Hexane		
Cyclohexane		
Ethanol		
Acetonitrile		
Methylene chloride		

2. Is there a relationship between a solvent's polarity and the wavelength of its absorbance maximum?

3. Plot the UV absorption spectra of the sample in the different solvents on the same graph.

Experiment 15
Effect of Substituent with Unshared Electrons on UV-VIS spectra of compounds

Introduction

Among the factors affecting the wavelength maxima of compound during UV-VIS absorption is conjugation extension from lone pair electrons on substituents. The non-bonding electrons increase the length of π-system through conjugation and shift the primary and secondary absorption bands to longer wavelength. The more the availability of these non-bonding electrons, the greater a bathochromic shift shift will take place due to extended conjugation. In addition, the presence of non-bonding electrons introduces the possibility of n $\rightarrow$ π^* transitions. If the non-bonding electron is excited into the extended π^*chromophore, the atom from which it is removed becomes electron-deficient and the π-system of aromatic ring becomes electron rich. This situation causes a separation of charge in the molecule and such excited state is called a charge-transfer or an electron-transfer excited state.

In going from benzene to t-butylphenol, for example, the primary absorption band at 203.5 nm shifts to 220 nm and secondary absorption band at 254 nm shifts to 275 nm. Further, the increased availability of n electrons in negatively charged t-butylphenoxide ion shifts the primary band from 203.5 to 236 nm (a 32.5 nm shift) and secondary band shifts from 254 nm to 290 nm (a 36 nm shift) (Figure 1 below). Both bands show hyperchromic effect. On the other hand, in the case of anilinium cation, there are no n electrons for interaction and absorption properties are quite close to benzene. But in aniline, the primary band is shifted to 232 nm from 204

70

nm in anilinium cation and the secondary band is shifted to 285 nm from 254 nm (Figure 2).

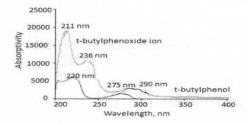

Figure 1: *UV-spectra of t-butyl phenol and t-buty phenoxide in methanol*

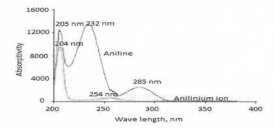

Figure 2: *UV-spectra of aniline and anilinium salt in methanol*

Chemicals and Apparatus

o   0.01 % (w/v) aqueous solution of sulphanilamide

o   1 N hydrochloric acid, 1 N sodium hydroxide

Purpose

In this experiment we will effect of presence or absence of non-bonding electrons on wavelength of absorption maxima of sulphanilamide solution. In alkaline solution, the amino group is retained as a chromophore while in acidic medium it is converted to quaternary ammonium ion loosing the non-bonding electrons.

$$H_2N\text{-}C_6H_4\text{-}SO_2\text{-}NH_2 + H^+ \rightarrow {}^+H_3N\text{-}C_6H_4\text{-}SO_2\text{-}NH_2$$

Procedure

1. Pipet 10.0 mL of sulphanilamide solution into 100 mL standard flask, dilute to the mark with 1N hydrochloric acid solution and mix. Determine the absorbance of this solution in 1 cm cells at 10 nm interval from 210-300 nm, using 1 N hydrochloric acid as the blank (use 5 nm interval around λmax)·

71

2. Repeat step (1) but using 1N sodium hydroxide instead of 1N hydrochloric acid,

3. Plot the absorption curves of the acid and alkaline solutions on the same graph paper.

4. Determine λ_{max} and calculate ε_{max} for each solution.

Sulfanilamide in	Data Table: Absorbance measurements at different λ(nm)								
	220	230	240	250	260	270	280	290	300
HCl									
NaOH									

Post-Lab

1. Discuss the effect of pH on absorption spectra of aniline and phenol.
2. Plot the absorption curves of the acidic and alkaline solutions of sulphanilamide on the same graph paper.

Sample Lab report
Experiment xx
Determination of percentage of ethanol in some brands of alcohols
Introduction

   The concentration of ethanol in alcoholic beverages is generally measured as percent alcohol by volume. However, many alcoholic beverages, such as whiskey and vodka, are not labeled with the percentage of alcohol. Rather, they are labeled with the proof value, which is twice the volume percentage of the alcohol in solution [1]. Thus, 80 proof whiskeys contain 40 percent ethanol. In this laboratory experiment, it was attempted to experimentally verify that the alcohol content of 80 proof Monarch brand whiskey was 40 percent by volume. The analysis of alcohol content in this experiment utilizes the density relationship, which relates the quantity of matter to the volume it occupies. The densities of many pure substances are known and tabulated. For example, at 20 ^{0}C the density of water is 1.00 g/mL and the density of ethanol is 0.789 g/mL [2]. The densities of mixtures, such as whiskey, reflect the components that make up the mixture. For example, the density of a mixture of water and ethanol would be expected to be less than the density of water and more than the density of ethanol. When the density is unknown, it can be determined by weighing a known volume of water on an analytical balance and calculating using the equation: *density = mass/volume.* For this laboratory, a series of ethanol/water solutions were

72

prepared by mixing known volumes of pure ethanol with known volumes of water. Aliquots of each mixture were then weighed and the density of each solution calculated. A plot of density vs. concentration (percent by volume) was prepared and the concentration of the whiskey was determined using the results of a linear regression analysis.

Objectives

Determination of percentage of ethanol in whisky

Calculation of density of whisky

Evaluating the accuracy of the measurements

Procedures

30%, 45%, 50%, 60%, 75%, 85%, and 95% ethanol/water solutions were prepared and 5.00 mL aliquots were weighed on a balance. The mass and calculated density of each solution was recorded as shown in table 1 below.

Table1: Mass and density of ethanol/water solutions

% ethanol(v/v)	30	45	50	60	75	85	95
mass (g)	4.750	4.470	4.581	4.505	4.280	4.191	4.000
density (g/mL)	0.950	0.950	0.916	0.901	0.856	0.838	0.800

Density calculation: *density = mass/volume=* 4.7505g/5.00mL = 0.950 g/mL

A 5-mL aliquot of white horse whiskey was weighed on balance and the density determined:

mass of whiskey = 4.710 g, and density of whiskey = 0.942 g/mL

Results and Discussion

A graph was prepared by plotting solution density vs. percentage by volume ethanol (table 1). A linear fit of the data was done where the equation of the best fit line was found to be: $y = -0.0021x + 1.0124$, or in other words, *density of the solution* $=(-0.0021)(percentage\ alcohol) + 1.0124$. The percentage alcohol in whiskey was determined by rearranging this equation and using the calculated density of whiskey:

percentage alcohol = (density of the solution -1.0124)/-0.0021 = 46%

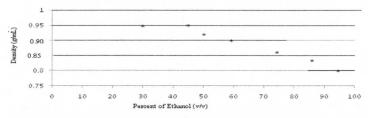

Figure xx: plot of density vs volume percentage of ethanol/water mixture

73

The determination of the concentration of the components of a mixture can often be done utilizing indirect methods. This experiment was performed for the purpose of determining the alcohol content (percentage by volume) of whiskey by measuring its density. Whiskey is a mixture of water, ethanol and various other substances that influence the flavor and color. The densities of pure water and pure ethanol are known, but when the two are combined, the density of the resulting solution is a function of the ratio of the quantities of the two substances. Monarch brand whiskey was determined experimentally to have a density of 0.942 g/mL, which correlated to 46% ethanol by volume. This result contrasts somewhat with the 40% value stated on the bottle.

Conclusion

Before jumping to the conclusion that the bottle had been mislabeled, it is important to consider many sources of error that were present in the experimental procedure. The first to note is that density is a function of temperature and temperature recordings were not made during this procedure. It was assumed that the temperature remained constant but there may have been some small fluctuations that would contribute to indeterminate error. There was also indeterminate error present due to the use of the volumetric pipettes. As this was a first experience with a pipette, the technique was somewhat imprecise. In particular, the use of the graduated volumetric pipette in the preparation of the known solutions was a challenging experimental technique. Too much ethanol added would cause a density calculation to be low, while too little ethanol added would cause a high density calculation. These two sources of indeterminate error were the primary sources of the scatter in the data, clearly seen in the graph.

Table 1 and graph 1 also reveal that there is a problem with the data recorded for the 45% by volume solution. I am not sure whether the solution was prepared improperly or the density was measured improperly, but the non-linearity of this data point suggests that the measurements should have been repeated if more time had been available. If this point is eliminated, the calculated value for whiskey drops to 44% by volume. Sources of determinate error in this procedure could be found in the calibration of the pipets and balances. Given the magnitude of the indeterminate error described above, it should not be believed that determinate errors in the volume measurements were significant. However, different balances were used to weigh the known solutions and the whiskey sample. Inconsistencies in the balances would contribute to determinate error that could be eliminated by using the same balance.

74

There is also a flaw in the basic premise of this experiment. That is, whiskey with a large number of "impurities" was compared to mixtures of pure ethanol and water. Surely the other substances present in whiskey will affect its density to some extent and therefore also affect the percent by volume determination. However, this experiment was still a useful exercise in applying the concept of density to a real world question.

References

1. Miller, G. Tyler; *Chemistry: Principles and Applications*, Wadsworth, Belmont, CA; **1976.**

2. Petrucci, Ralph; *General Chemistry, 5/e*; Macmillan, N.Y., N.Y.; **1989**.

Appendix 1: Sections of a lab report

Title	Specific, clearly conveys purpose of lab, can be abbreviated on subsequent pages.
Date	On first page, original date of starting lab activities
Lab Partners	Clearly listed on first page of lab report
Purpose	One sentence or two explaining the purpose of the experiment or the problem being investigated, Be specific but concise, this should relate directly to the conclusion you will draw, you may want to add to or change your purpose after completing the lab.
Materials	Complete list in columns or bullets
Procedure	Written as a list of numbered steps. Steps taken are specific enough so that someone not familiar with the lab or your work could do the procedure and repeat your results, Changes to procedure within a trial can be documented in observations, and you may need to include safety procedures if any.
Data Tables	Make table(s) large enough to write in, You may have to create your own table(s) if one is not given on your manual before coming to class.

Appendix 2: A marking guide (rubric) for instructors

Heading	Criteria	Performance level	points
Aim/objective	Purpose of the experiment stated in one's own words using clear, simple sentences	No ---------- Good ---------- Very good ---------	0 ½ 1
Introduction	Conceptuality, relevance to topic significance, language usage, clarity citation available	Fair ---------- Good ---------- Very good ---------	½ 1 2
Methods/ Procedures	Detailed steps written in passive voice Methods of data analysis included Relationships between dependant and independent variables indicated	Fair ---------- Good ---------- Very good ---------	 ½ 1 2
Results	Data collected in table formats Graphs are available Graphs and tables are labeled well	Fair ---------- Good ---------- Very good ---------	½ 1 2
Discuss ion	Chemical equations, if any Calculations done properly Discussion if the results agree with theory or hypothesis Any possible sources of errors discussed Any attempt to reduce error indicated	Fair ---------- Good ---------- Very good ---------	 ½ 1 2
Over all Lab report structure	Cover page style Neatness and readability Tables and graphs have title Pages are numbered	Good Very good	½ 1
		Total points	10

References

1. D.A. Skoog and J.J. Leary, Principles of Instrumental Analysis, 4 [th] Ed. Saunders College Publishing, 1992.

2. D.A. Skoog, D.M. West and F.J. Holler, Fundamentals of Analytical Chemisgtry, 2004.

3. G.D. Christian, Analytical Chemistry, 5th Ed., John Willey and sons, Inc., New York, 1994.

4. Wang, Analytical Electrochemistry, 2nd Ed. John Wlley & sons Ltd, 2001

5. Chemistry: matter and change, Laboratory manual, Glencoe McGraw-Hill Company, student edition

6. www.chemistry.sc.chula.ac.th/bsac/Org%20Chem%20Lab.../Exp.3[1].pd....

7.http://www.gobookee.org/experiment-15-paper-chromatography-lab-report-conclusion/

8. chemistry.syr.edu/totah/che276/support/5a1.exp/1.tlc4.pdf

9. www.vrml.k12.la.us/rpautz/documents/.../conductometrictitration.pdf

10. www.fpharm.uniba.sk/.../Determination_of_the_specific_conductance.p...

11. www.nitm.ac.in/Documents/students/Expt%206.pdf

12. www.tau.ac.il/~chemlaba/Files/conductometry-titrations.pdf

13. chemweb.chem.uconn.edu/teaching/.../GA7_potentio_Titr_rev7_99.pdf

14. www.asso-etud.unige.ch/aecb/rapports/3eme/.../titration_08.pdf

15. http://mason.gmu.edu/~jschorni/chem211lab/Chem%20211-212%20Vinegar.pdf

16. http://www.profpaz.com/Files/chem52/Exp_5.pdf

17. http://www.chemcollective.org/chem/sinclair/exp9virtual2.pdf

18. http://faculty.lacitycollege.edu/boanta/LAB101/2013/titration13.pdf

19. http://www.chemcollective.org/chem/sinclair/exp9virtual2.pdf

20. http://www.csun.edu/~jeloranta/CHEM355L/experiment4.pdf

21.http://www.phywe.com/index.php/fuseaction/download/lrn_file/versuchsanleitung en/P30606,_60/e/P3060660e.pdf

22. John R. Dean, Extraction Techniques in Analytical sciences, Northumbria University, Newcastle, UK, Wiley and Sons publications, 2009

23. Agilent Technologies, Fundamentals of UV-VIS spectroscopy, workbook, 2000

24. Robert D. Braun, introduction to chemical analysis, McGraw-Hill Book Company, 1985

25. Determination of Iodate in Iodized Salt by Redox Titration, Outreach, College of Science, University of Canterbury, Private Bag 4800, ChristchurchNew Zealand

26. Bruchertseifer H, Cripps R, Guentay S, Jaeckel B, Analysis of iodine species in aqueous solutions. Anal Bioanal Chem 375:1107–1110, 2003

27. Bürgi H, Schaffner T, Seiler JP (2001) the toxicology of iodate: a review of the literature. Thyroid 11:449–456

28. Hetzel BS, Iodine deficiency disorders (IDD) and their eradication. Lancet 2:1126– 1127, 1983

29. Kulkarni et al, A rapid assessment method for determination of iodate in table salt samples, Journal of Analytical Science and Technology, 4:21, 2013

30. Visser TJ The elemental importance of sufficient iodine intake: a trace is not enough. Endocrinol 147:2095–2097, 2006

31. Zimmermann MB, Iodine deficiency. Endocrine Rev 30:376–408, 2009